AF787003

ION IMPLANTATION - RESEARCH AND APPLICATION

Edited by **Ishaq Ahmad**

Ion Implantation - Research and Application
http://dx.doi.org/10.5772/65528
Edited by Ishaq Ahmad

Contributors

Ghulam Husnain, Morgan Madhuku, Dongli Fan, Xin Zhou, Yiming Zhang, Xiaohua Shi, Heriberto Marquez, Gloria Veronica Vazquez, Alicia Oliver, Eder Lizarraga, Raul Rangel-Rojo, David Salazar, Sonia Kaschieva, Sergey N. Dmitriev, Yanina Fedorenko, Ishaq Ahmad

Published by InTech
Janeza Trdine 9, 51000 Rijeka, Croatia

© The Editor(s) and the Author(s) 2017
The moral rights of the editor(s) and the author(s) have been asserted.
All rights to the book as a whole are reserved by InTech. The book as a whole (compilation) cannot be reproduced, distributed or used for commercial or non-commercial purposes without InTech's written permission. Enquiries concerning the use of the book should be directed to InTech's rights and permissions department (permissions@intechopen.com).
Violations are liable to prosecution under the governing Copyright Law.

Individual chapters of this publication are distributed under the terms of the Creative Commons Attribution 3.0 Unported License which permits commercial use, distribution and reproduction of the individual chapters, provided the original author(s) and source publication are appropriately acknowledged. More details and guidelines concerning content reuse and adaptation can be found at http://www.intechopen.com/copyright-policy.html.

Notice

Statements and opinions expressed in the chapters are these of the individual contributors and not necessarily those of the editors or publisher. No responsibility is accepted for the accuracy of information contained in the published chapters. The publisher assumes no responsibility for any damage or injury to persons or property arising out of the use of any materials, instructions, methods or ideas contained in the book.

Publishing Process Manager Marijana Francetic
Technical Editor SPi Global
Cover InTech Design team

First published June, 2017
Printed in Croatia
Legal deposit, Croatia: National and University Library in Zagreb

Additional hard copies can be obtained from orders@intechopen.com

Ion Implantation - Research and Application, Edited by Ishaq Ahmad
p. cm.
Print ISBN 978-953-51-3237-0
Online ISBN 978-953-51-3238-7

We are IntechOpen,
the world's leading publisher of
Open Access books
Built by scientists, for scientists

3,650+
Open access books available

114,000+
International authors and editors

119M+
Downloads

151
Countries delivered to

Our authors are among the
Top 1%
most cited scientists

12.2%
Contributors from top 500 universities

WEB OF SCIENCE™

Selection of our books indexed in the Book Citation Index
in Web of Science™ Core Collection (BKCI)

Interested in publishing with us?
Contact book.department@intechopen.com

Numbers displayed above are based on latest data collected.
For more information visit www.intechopen.com

For EU product safety concerns:
IN TECH d.o.o., Prolaz Marije Krucifikse Kozulić 3, 51000 Rijeka, Croatia,
info@intechopen.com or visit our website at intechopen.com.

Meet the editor

Ishaq Ahmad, PhD, is the director of Experimental Physics of NCP (National Centre for Physics), Islamabad, Pakistan. He has coauthored over 80 publications and supervised several master's, PhD, and postdoctoral degree students. His research interests are focused on ion implantation, ion beam–induced modification of materials/nanomaterials, ion beam–induced welding of nanowires/nanotubes, and ion beam analysis of materials. He is an editorial board member and a reviewer of several well-known journals. He was a TWAS-UNESCO-iThemba LABS research associate from 2014 to 2016. He is a senior fellow of UNESCO UNISA AFRICA chair in nanosciences/nanotechnology and senior fellow of the Nanosciences African Network (NANOAFNET), an ICTP network.

Contents

Preface

Ion implantation is a versatile technique broadly implemented in different areas of science and technologies. It has been observed as a continuously evolving technology. Ion Implantation is a reliable technology due to its accurate control over the dopant compositions and penetration depth through the choice of the species and the energies of the ions. Indeed, importance of ion implantation technology has been recognized to produce many commercial products. However, the development of new devices and materials through this technology is still looking for the benefit of human being.

In addition, MeV ion implantation and defects engineering are another new emerging science and technology class where research and innovation are going on. As for as the materials science concern, defects play an important role in the pattern of properties of materials that may be tuned according to our needs, whereas defects engineering in biocompatible materials is a promising technology for enhancing the properties of biomaterials. Surface modifications of polymers for the improvement in compatibility of the blood and tissue are few other examples where the ion implantation is playing a vital role.

In this book, there is a detailed overview of the recent ion implantation research and innovation along with the existing ion implantation technological issues especially in microelectronics. The book also reviews the basic knowledge of the radiation-induced defects production during the ion implantation in case of a semiconductor structure for fabrication and development of the required perfect microelectronic devices. The improvement of the biocompatibility of biomaterials by ion implantation, which is a hot research topic, has been summarized in the book as well. Moreover, advanced materials characterization techniques are also covered in this book to evaluate the ion implantation impact on the materials.

The substantial part of this book concise the influence of ions on the characteristics of the material in particular semiconductor materials and devices. The third section of this book relates MeV ion beam appointment to fabricate devices. The final section of the book delineates the impact of the ion beam on the properties of biomaterials.

Finally and foremost, I wish to record my sincere appreciation to the researchers who potentially contributed in this book undertaking: Prof. Dr. Y. G. Fedorenko, Prof. Dr. Heriberto Márquez, Prof. Dr. Gloria V. Vázquez, Prof. Dr. Eder G. Lizárraga-Medina, Prof. Dr. Raúl Rangel-Rojo, Prof. Dr. David Salazar, Prof. Dr. Alicia Oliver, Prof. Dr. Xin Zhou, Prof. Dr. Yiming Zhang, Prof. Dr. Xiaohua Shi, Prof. Dr. Dongli Fan, Prof. Dr. S. Kaschieva, Prof. Dr. S. N. Dmitriev, Prof. Dr. G. Husnain, Prof. Dr. Morgan Madhuku, Dr. Shehla Honey, and Ms. Marijana Francetic, Publishing Process Manager.

Ishaq Ahmad
National Centre for Physics
Islamabad, Pakistan

Section 1

Introduction

Chapter 1

Introductory Chapter: Introduction to Ion Implantation

Ishaq Ahmad and Waheed Akram

Additional information is available at the end of the chapter

http://dx.doi.org/10.5772/intechopen.68785

1. Introduction

This chapter elucidates the concept of low-energy/high-energy ion implantation and its key applications in materials science. Ion implantation is the interaction of energetic ion beam with solids. In this ion-solid interaction, ions penetrate through the materials and slow down to some extent into the materials due to electronic and nuclear energy loses. From the 1950s to a few decades back, ion beam implantation has only been known as a process used for damaging the surface of bulk materials and ion implantation of semiconductors to make p-type or n-type materials. However, ion implantation has been proven in recent studies as a reliable technique to tune the properties of bulk materials, thin films, nanostructure materials and biocompatible materials for specific applications [1–4]. Nevertheless, material properties can be altered as per proper selection of ion species, ion energy, substrate temperature, and ion fluencies.

Energetic ions usually consider here in this chapter ranging from keV to hundreds of MeV are mainly produced from different ion sources. These ions are then accelerated up to required energies according to their applications. For surface treatment of solids or ions doping in semiconductors, usually low-energy ion implanters in keV ranges are applied. Recently developed FIB is categorized as low-energy ion system, which enables heavy ion micromachining to fabricate micro and nanodevices. Whereas, medium-energy ions from medium energy electrostatic accelerators are used for proton beam writing, synthesis and modification of thin films [5, 6]. High-energy protons and Swift's heavy ions having hundreds of MeV energies are applied for a surface modification and intrinsic physical properties of thin films [7]. High-energy protons and Swift's heavy ions can be produced normally in electrostatic accelerators or cyclotrons.

For reader's convenience, further elaboration of the ion energies into three broad categories and their impacts on materials is as follows:

INTECH
open science | open minds

© 2017 The Author(s). Licensee InTech. This chapter is distributed under the terms of the Creative Commons Attribution License (http://creativecommons.org/licenses/by/3.0), which permits unrestricted use, distribution, and reproduction in any medium, provided the original work is properly cited.

1.1. Low-energy ion implantation: range ~ 1 to 200 keV

Ion implantation is usually the low-energy process to introduce doping atoms into a semiconductor wafer to form devices and integrated circuits. Low-energy ion implanter is shown in **Figure 1**. In low-energy ion implanter system, ions of materials are generated and accelerated through the electric field and then irradiate on samples.

Presently, advanced applications of low-energy ion implanters include modification of the physical, chemical, or electrical or magnetic properties of thin films and nanostructure materials through doping of atoms as well as defect production. Sometimes, synthesis of doped nanostructure materials is difficult through chemical methods but ion implantation made it possible to dope required atoms. Recently, Ishaq et al. used ion implanter to irradiate carbon nanostructures with 70 keV H, N, and Ar ions to change the morphology and structure [8]. Moreover, low-energy ion implanter can be used to weld nanowires, nanotubes, or integrate nanowires to make nanodevices, which are a unique application of ion implanter in nanotechnology [9, 10]. Low-energy ion implanter was utilized to weld carbon nanotubes and fabricate carbon nanotubes network by ion beam irradiation to improve electrical properties as shown in **Figure 2** [9, 11]. In addition, low-energy ion implanter was utilized to make metal-semiconductor junctions for future device fabrication [10]. In low-energy regime, ions interact with nanowires deposit energy to target atoms and materials are sputtered. These sputtered atoms deposit on junction positions and same time ion beam induced collision cascade effects to weld nanowires/nanotubes.

Recently, a low-energy FIB system has been developed for controlled three-dimensional (3D) micromachining and fabricates ultra-modern micro and nanodevices used in different applications [12]. Either this system can be used for precise doping in nanoscale regime or implant in few atoms in biological samples for DNA damage studies. Low-energy FIB is also used in deposition and ablation of materials. In this FIB system, usually heavy ions such as

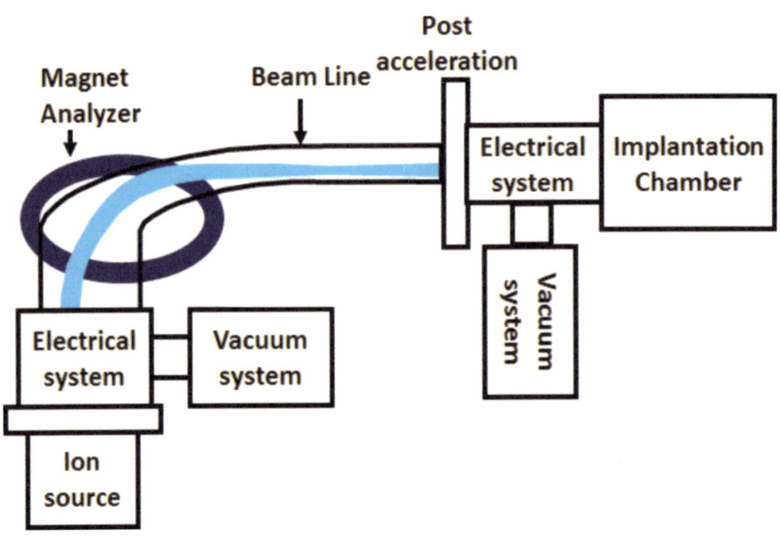

Figure 1. Ion implanter.

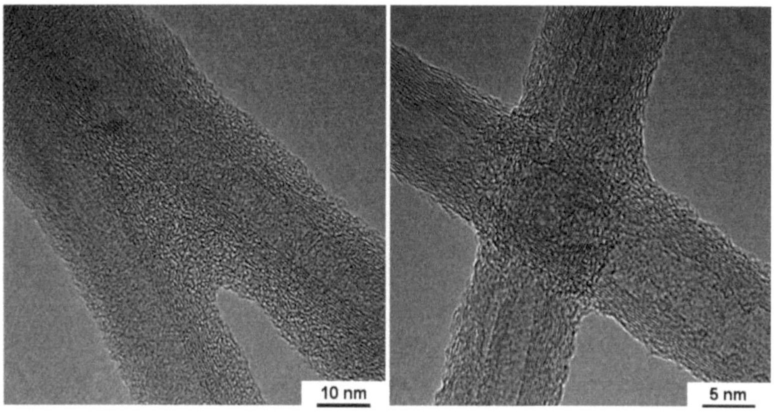

Figure 2. Welding of carbon nanotubes by keV ion implantation [11].

Ga ions hit the target materials and sputter atoms from the target samples due to elastic collision, allowing precision milling of the specimen down to a submicrometer or even a nanoscale. More applications of low-energy ion implanter are presented in the proceeding chapters of this book.

1.2. Medium-energy ion implantation: range ~ 300 keV to 50 MeV

Medium-energy ions are usually generated and accelerated from medium-energy ion beam accelerators such as Van de Graaff or pelletron accelerators. Due to the advancement in accelerator technologies and instrumentations, we can get controlled ion beam with a different spot size of the ion beam. Even nowadays, single atom can be accelerated and implanted into the required position of samples. In medium-energy ion beam accelerators, one or two types of different ion sources are attached, which generate almost all types of ions from hydrogen to uranium. These ions are then accelerator through high-energy system as shown in **Figure 3**. Micro beamline can be used for proton beam writing where micro and nanodevices can be fabricated, whereas ion implantation beamline can be used for medium-energy ion implantation into the materials.

Medium-energy ion accelerators are versatile and advanced technologies can be applied in the different field of sciences. Regarding ion implantation, medium-energy ions are effectively

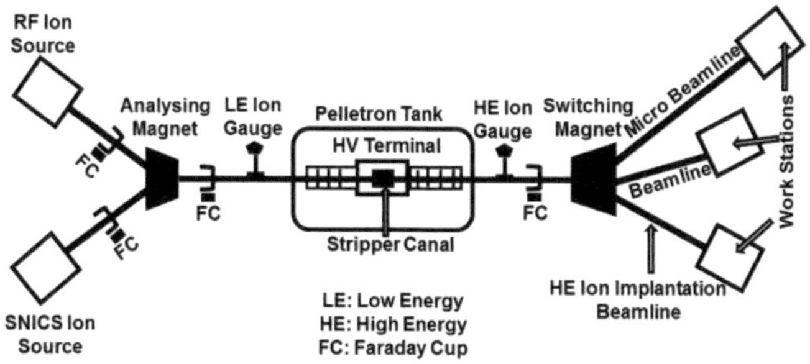

Figure 3. Tandem pelletron accelerator.

utilized for deep ion implantation purpose. It is possible to implant required ion species into required depth of samples precisely. High-energy ions have greater penetrating capabilities in materials while maintaining a straight path.

During ion implantation, ion beam induced collision cascade effect and induced surrounding heat along the path of ions tracks, which should be managed for proper ion implantation to prevent damage into the target material. In the case of uncontrolled ion beam irradiation or high current irradiation, induced ion beam produced great defects, which lead the target materials to amorphization or phase transformation. Therefore, low implantation ion beam current is advised to minimize ion beam induced local heating for prevention of amorphization or other phase transformations. Ishaq et al. successfully implanted C-atoms into BNNTs using MeV C ions through pelletron accelerator to form boron carbonitride nanotubes [13]. Whereas uncontrolled MeV ion beam implantation on crystalline silver nanowires lead to form amorphous silver nanowires [14]. Sometimes, high-energy ion implantation is required to produce defects into materials for functionalization of materials especially polymer-based materials for biological applications or enhance the absorption properties of materials [4]. For this purpose, the high current ion beam is required to create such defective structures. Additionally, these defective structures are important for different applications such as attached functional groups and enhanced sensing properties of gas sensors etc. These medium-energy ions are also explored new application in nanotechnology. Recently, tandem accelerator was employed to weld nanowires or making large area welded network of nanowires [15]. High-energy ions produce local heating along the path of ion track and at the same time collision cascade effects make a rearrangement of atoms at the junction point which results in welding of metal nanowires. Recently, Shehla et al. weld silver nanowires by medium-energy ion beam implantation [16]. More applications of medium-energy ion implanter are well presented in the proceeding chapters of this book.

1.3. High-energy ion irradiation: range ~ 50 MeV to hundreds of MeV

High-energy ions include protons and swift heavy ions (SHI) are usually generated and accelerated from high-energy ion beam accelerators such as a cyclotron or high potential terminal voltage tandem electrostatic accelerators, same as shown in **Figure 3**. Moreover, cyclotron produces high-energy ions with high ion currents of the order of a milliamperes, which is useful for many applications. High-energy protons, alpha particles, and deuterons are used for radioisotopes production from stable elements for medical applications. The application of SHI includes characterization of materials and modification of materials to radioisotopes production for medical treatments, etc. Atomic displacements caused by SHI irradiation produces collision cascade effects which allow the target material to modify its properties. Additionally, SHI also helps to produce structural defects in materials to change the chemical, optical, electrical, or magnetic properties. Ion beam mixing induced by SHI irradiation is another application where some thin film alloys are difficult to fabricate. Through SHI ion beam mixing, such alloys are now possible. For example, TiBe alloy thin film is difficult to fabricate through chemical or physical processes. In ion beam mixing, just make multilayer Ti and Be films through thin film coating system and irradiate SHI to mix atoms. TiBe alloy will perfectly fabricate due to collision cascade effects, electronic excitation, and ion beam induced local heating along the ion track.

This book covered various topics regarding the use of different ion energies, ion beam accelerators, and impacts of ions application in materials science studies. The impact of these ions on materials is characterized using long range of advanced analytical techniques, such as nuclear reaction analysis (NRA), elastic recoil detection analysis (ERDA), Rutherford backscattering spectrometry (RBS), RBS/channeling, high-resolution X-ray diffraction (HRXRD), XRD, alternating gradient magnetometer (AGM), SQUID magnetometer, positron annihilation spectroscopy (PAS), positron annihilation–induced Auger electron spectroscopy (PAES), electron spin resonance spectroscopy, low-high frequency CV measurements, deep level transient spectroscopy (DLTS), internal photoemission spectroscopy (IPE), prism coupling technique for refractive index measurement, thermally stimulated current (TSCM), soft X-ray emission spectroscopy (SXES), Fourier transform infrared spectroscopy (FTIR), scanning electron microscope (SEM), X-ray photoelectron spectroscopy (XPS), zeta potential analyzer, atomic-force microscopy (AFM), and environment control scanning probe microscope.

Author details

Ishaq Ahmad* and Waheed Akram

*Address all correspondence to: ishaq.ahmad@ncp.edu.pk

National Center for Physics, Islamabad, Pakistan

References

[1] Husnain G, Yao S-D, Ishaq A, Rafique HM. Structural and magnetic impact of Cr^+ implantation into GaN thin film. Solid State Sciences. 2012;**14**:735-738

[2] Ni Z, Ishaq A, Yan L, Gong J, Zhu D. Enhanced electron field emission of carbon nanotubes by Si ion beam irradiation. Journal of Physics D Applied Physics. 2009;**42**:075408.

[3] Ishaq A, Yan L, Zhu D. The electrical conductivity of carbon nanotube sheets by ion beam irradiation. Nuclear Instruments and Methods in Physics Research Section B: Beam Interactions with Materials and Atoms. 2009;**267**:1779-1782

[4] Tsuji H, Sommani P, Kitamura T, Hattori M, Sato H, Gotoh Y, Ishikawa J. Nerve-cell attachment properties of polystyrene and silicone rubber modified by carbon negative-ion implantation. Surface and Coatings Technology. 2007;**201**:8123-8126

[5] Shehla H, Ishaq A, Awais A, Thema FT, Javed I, Wang D, Rauf K, Nasir K, Naseem S, Maaza M. Silver nanowires stability and burying into substrates under MeV proton irradiation. Current Nanoscience. 2016;**12**:774-780

[6] Shakil K, Ishaq A, Noaman K, Mazhar M, Waheed A, Maaza M. Carbon ions irradiation induced modifications in structural and electrical resistivity characteristics of ZrN thin films. Materials Science in Semiconductor Processing. 2015;**39**:530-535.

[7] Smita S, Narayan H, Sulania I, Thakurdesai M. Swift heavy ion irradiation induced nanograin formation in CdTe thin films. Nuclear Instruments and Methods in Physics Research Section B: Beam Interactions with Materials and Atoms. 2016;**387**:1-9

[8] Ishaq A, Shahid I, Naveed A, Khurram AA, Akrajas AU, Dee CF, Naseem S, Rafique HM, Long Y. H^+, N^+, and Ar^+ ion irradiation induced structure changes of carbon nanostructures. New Carbon Materials. 2013;**28**:81-86

[9] Ishaq A, Ni Z, Yan L, Gong J, Zhu D. Constructing carbon nanotube junctions by Ar ion beam irradiation. Radiation Physics and Chemistry. 2010;**79**:687-691

[10] Dee CF, Ishaq A, Long Y, Zhou X, Majlis BY. Contact welding study of carbon nanotube with ZnO nanowire. Physica E: Low-Dimensional Systems and Nanostructures. 2011;**43**:1857-1862

[11] Yan L, Zhou G, Ishaq A, Zhou X. Improving the electrical conductivity of multi-walled carbon nanotube networks by H ion beam irradiation. Carbon. 2011;**49**:2141-2144

[12] Kim SJ, Yamashita T, Nagao M, Sato M, Maeda H. Development of 3D focused-ion-beam (FIB) etching methods for fabricating micro- and nanodevices. The Japanese Journal of Applied Physics. 2002;**41**:4283-4286

[13] Ishaq A, Naqvi SR, Iqbal J, Bo L, Yan L, Dee CF, Baig A. Substitutional carbon doping of hexagonal multiwalled boron nitride nanotubes (h-MWBNNTs) via ion implantation. The Journal of Nanoparticle Research. 2014;**16**:2170-2178

[14] Shehla H, Awais A, Zongo S, Javed I, Ishaq A, Khizar H, Naseem S, Maaza M. Fabrication of amorphous silver nanowires by helium ion beam irradiation. Chinese Physics Letters. 2015;**32**:096101

[15] Shehla H, Naseem S, Ishaq A, Maaza M, Bhatti MT, Wan D. Large scale silver nanowires network fabricated by MeV hydrogen (H^+) ion beam irradiation. Chinese Physics B. 2016;**25**:046105

[16] Shehla H, Ishaq A, Yaqoob K, Javed I, Saira R, Shahzad N, Maaza M. Ion beam irradiation-induced nano-welding of Ag nanowires. Micro and Nano Letters. 2016;**11**:34-37

Section 2

Materials and Characterization

Chapter 2

Metal Ions Implantation-Induced Effects in GaN Thin Films

Ghulam Husnain and Morgan Madhuku

Additional information is available at the end of the chapter

http://dx.doi.org/10.5772/68042

Abstract

MOCVD-grown GaN n-type epilayers were implanted with 150keV Co^+ and Cr^+ ions at different fluences at room temperature. Co^+ was implanted at $3x10^{16}$ and 5×10^{16} ions/cm^2 and samples rapid-thermal-annealed at 700, 800 and 900°C for 5 minutes, while Cr^+ was implanted at $3x10^{16}$ ions/cm^2 and annealed at 800 and 900°C for 2 minutes. Diffraction patterns of implanted samples showed satellite peaks at the lower side of the main GaN (0002) reflection and these were assigned to implantation induced-damage and the formation of $Ga_{1-x}Co_xN$ or $Ga_{1-x}Cr_xN$ phases. The coercivity (H_c) at 5K from SQUID for Co^+ implanted GaN at $3x10^{16}$ ions/cm^2 was 275 Oe and that at $5x10^{16}$ ions/cm^2 was 600 Oe. For Cr^+ implanted GaN at $3x10^{16}$ ions/cm^2, Hc was 175 Oe. At the same dose of Cr^+ and Co^+ implanted ions, the saturation magnetization (*Ms*) values were almost similar. But after annealing at 900°C, the *Ms* value of Cr^+ implanted GaN was higher than that of Co^+ implanted at 5K. For Co^+ implanted GaN, magnetization was retained up to 370K while in Cr^+ implanted GaN, magnetization was retained above 380K. These findings are the highest reported Curie temperatures for Co^+ and Cr^+ implanted GaN diluted magnetic semiconductors.

Keywords: GaN, quaternary alloys, microstructure, ion implantation, dilute magnetic semiconductors

1. Introduction

III-Nitrides are currently finding applications in conventional devices, such as UV-Vis laser diodes, ultra-bright LEDs, UV detectors, high temperature electronics, high-density optical data storage, aerospace and automobiles technologies [1]. Furthermore,

INTECH
open science | open minds

© 2017 The Author(s). Licensee InTech. This chapter is distributed under the terms of the Creative Commons Attribution License (http://creativecommons.org/licenses/by/3.0), which permits unrestricted use, distribution, and reproduction in any medium, provided the original work is properly cited.

III-nitrides, when doped with magnetic impurities, have plentiful possibilities of being used in diluted magnetic semiconductors (DMSs) for spintronic device applications. Most of the above-mentioned devices require well-organized, controlled and targeted area doping. Even though materials can be doped during growth, ion implantation presents various advantages, which are not achievable by doping during growth in III–V nitrides. The low solubility of transition metals in III–V nitrides ($<10^{18}cm^{-3}$) has restricted their doping to produce a range of magnetic semiconducting materials. Major changes in magnetic properties of these materials are not expected because of limit in solubility since there is a direct relationship between magnetic effects and concentration of magnetic impurities.

Ion implantation has a number of advantages [2], some of which are that any dopant atom can be introduced at any desired depth and concentration above solid solubility limit in a material and that doping can be done on a precisely defined area [3]. Moreover, ion implantation offers other advantages such as electrical isolation, dry etching, quantum well intermixing and ion cut [4]. However, ion implantation has its own drawbacks such as lattice damage and generation of new defects [5]. An entrenched solution to get rid of implantation-induced damage is thermal annealing. Moreover, dopants can be optically, electrically and magnetically activated through thermal annealing. However, annealing conditions (time, temperature and atmosphere), sample thickness, cap layers and implantation parameters have to be considered to successfully optimize thermal annealing to recover the lattice.

Considerable progress has been made in micro-electronics since the time discrete circuit elements were replaced by integrated circuits. Scientists have shown keen interest to harness the spin of electrons so as to further improve the functionality of devices [6]. Currently, charge-flow is used to carry information between microelectronic devices; however, spin movement can also be employed to carry information [7]. This has provided prospects of utilizing charge and spin degrees of freedom concurrently to bring about a new generation of electronic devices known as spin-electronics or spintronics. Spintronics is a multidisciplinary field covering physics, chemistry and engineering in which electronic, opto-electronic and magneto-electronic features can be integrated on a single chip [8]. The search for new spintronics materials as well as ways to improve existing materials is still on-going. Diluted magnetic semiconductors (DMSs), which are a class of materials in which a small quantity of magnetic ions is introduced into normal semiconductors, have been found to be suitable for spintronic device applications [9].

Recent II–VI and III–V DMS doped with magnetic ions such as Mn, Fe, Co and Ni are (CdMn)Te, (GaMn)As, (InMn)As, (GaMn)Sb, (ZnMn)O and $(TiMn)O_2$ [10]. The most studied DMS are (GaMn)As and (InMn)As but these are limited by Curie temperature and are unsuitable in practical spintronic devices [11]. In 2000, Dietl et al. predicted that a Curie temperature above room temperature in GaN-based DMS was possible and this has rejuvenated DMS research [12]. The successful applications of III-V nitrides in electronic and photonic devices have energized researchers to explore the potential of these materials in spintronics. The search is still on-going, since the announcement of new DMS materials is continuously being seen in research journals and others.

Group III–V semiconducting materials, especially GaN, has attracted momentous attention because of the prediction of a T_C higher than room temperature [12] for (Ga,Mn)N doped with Mn (5 at.%). Many studies have observed ferromagnetic-like behaviour close to or higher than room temperature for (Ga,Mn)N [13, 14] and as high as 940 K by Sonoda et al. [15]. Theoretical calculations, based on local spin-density approximation, which assumed that Ga atoms were randomly substituted by magnetic atoms, have predicted diverse magnetic properties extending from spin-glass-like to ferromagnetic-like behaviour for GaN together with various concentrations of Cr, Co, Fe, Mn, Ni and V [16].

Nevertheless, GaN-based DMS with other transition metals such as Cr, Co and Ni have not been adequately investigated. Above room temperature, ferromagnetism has been observed in TiO [17] and ZnO [18] doped with cobalt ions. The implantation of various semiconductors with magnetic ions in the search for possible DMS systems has been found to be effective [19]. A few recent studies on electrical, magnetic and optical properties of cobalt ion-implanted GaN [20, 21] and ZnO [22, 23] films have been reported. However, there are not many reported experimental studies on Co^+ implanted GaN as a function of annealing temperature in the literature.

Transition metal (TM)-doped III-nitride semiconductor films are important in the emerging spintronic applications due to the observed room-temperature ferromagnetic properties [24–26]. Doping with transition metal ions also appears to be an interesting way of producing high-resistivity buffer layers in emerging III-nitride-based high electron mobility (HEMTs) transistors [27–29]. Transition metals can be introduced into group III-nitride films during growth by metal organic chemical vapour deposition (MOCVD) [30, 31]. The influence of transition metal (TM) impurities on the electrical properties of both n-GaN and p-GaN has been reported [32–34]. Hashimoto et al. [35] have grown epitaxial GaCrN films by electron-cyclotron-resonance molecular beam epitaxy (MBE). The films showed ferromagnetic behaviour with a Tc higher than 400 K. To the best of our knowledge, there is currently no information on the magnetization of Cr-implanted GaN, at high fluences, in the literature.

In this chapter, we present magnetic and structural properties of Co^+ and Cr^+ implanted GaN as a function of annealing temperature. Here, n-type GaN epilayers were grown on sapphire by metal organic chemical vapour deposition (MOCVD) and subsequently implanted with Co^+ and Cr^+ metal ions. The properties of Co^+ and Cr^+ implanted GaN epilayers were investigated by structural and magnetic techniques. Section 2 discusses ion implantation in semiconductors, listing some of the advantages of ion implantation compared to doping during growth; thermal annealing after ion implantation and introduces the structural and magnetic techniques used to characterize the ion implanted GaN epilayers. Section 3 briefly describes the methods and materials used and Section 4 describes, in detail, cobalt and chromium implantation in GaN epilayers.

2. Ion implantation in semiconductors

Ion implantation is a conventional doping technique for device applications. In comparison with some modern techniques for thin film growth such as MBE, the implantation process can readily be used for making selected-area contact regions for injection of spin-polarized current into device structure. The incorporation of desired atoms into semiconductor materials by ion implantation was first introduced by William Shockley. This technique was patented in U. S. in 1957 and the first commercial Ion-implanter was released on the market in the 1970s. **Figure 1** shows suitable elements in the Periodic table for implantation into semiconductors [36].

Through ion implantation, impurity ions are projected into the target material to modify its structural, optical and electronic properties. Ion implantation is a relatively simple process to introduce ions into semiconductors for doping, electrical isolation of active regions and device applications. Some of the advantages of doping by ion implantation over doping during growth are listed hereunder.

- It is a rapid means of achieving the needed fluence.
- Species of most types of ions can be introduced into essentially any host material.
- Ions can be projected to the required depth by controlling the implantation energy.

Figure 1. Periodic table elements used for ion implantation [16].

http://dx.doi.org/10.5772/68042

- A proper mask system can be used to implant ions into the desired areas of the sample.
- Desired concentrations of dopants can be implanted into the material by monitoring implantation dose which is measured by the ion current.
- Ion species can be introduced over the solid solubility limits.
- It is insensitive to the properties of the host material like sample geometry, lattice structure, lattice defects and so on.
- Doping by ion implantation is a low temperature process.

An ion implanter consists of an ion source, a mass spectrometer, high voltage accelerators, a scanning system and a target chamber. The desired implant species are usually in the form of a gas. However, vapour from a heated solid can also be used. To generate electrons, an electric potential close to 100 V is applied across the body of the chamber and the filament and the electrons emitted from the filament ionize gas atoms by impact. The exit slit is put to a potential of about –25 kV and hence positive ions are accelerated towards it. They then enter the mass spectrometer which selects ions according to their masses and charge states and allow desired ion species to pass through the exit slit of the spectrometer.

The selected ions are then accelerated by passing through the accelerating columns at high voltages. These ions pass through the pair of X and Y plates of a scanner system to produce a beam for uniform dose implantation. The scanned beam then enters the defining aperture and is projected onto the material to be implanted. Electrical contact between the target and holder allows the flow of electrons to neutralize the dopants received by the target, which are counted by the charge integrator by measuring the time averaged swept beam current according to the relation,

$$Q = \int_0^t \frac{I}{nqA}\,dt \tag{1}$$

where '*A*' is the target area, '*n*' is the charge state of the ion beam and '*I*' is the beam current summed over the duration of implantation. A small positive potential is used to bias the target to reduce errors from secondary electrons. The energy of ions accelerated towards the target depends on the applied electrostatic potential and is known as the implantation energy. The implantation energy is normally in the range 1–2000 keV, depending on the implanter set-up. The number of ions received by the target per unit area (ions cm^{-2}) is known as dose or fluence and is measured electronically by a charge integrator.

During Coulombic interactions with target atoms, the incident ion beam loses its energy through scattering. If the complications of many-body interactions are disregarded, the supposition of binary collisions can be used as a suitable approximation. The implanted ion collides with a target atom, which in turn recoils and collides with other lattice atoms to start a collision cascade. The penetrating ion beam will encounter point defects (vacancy, interstitial, substitutional, anti-site, etc.) in its path. The mechanisms through which an ion beam loses energy can be described in two ways, which are elastic or nuclear energy loss (Coulomb interaction) and inelastic or electronic energy loss (electron-electron energy transfer, excitation and ionization phenomenon) [1]. The total distance travelled by the incident ion normal to the target surface before coming to rest is called range. It is generally expressed by symbol R and is given by

$$R = \int_0^E \frac{dE}{NS(E)} \tag{2}$$

where N is the atomic density, dE is the ion energy loss and $S(E)$ is the stopping cross-section. The projected range is an important parameter and calculated by the penetration depth of the ions below the target surface. This is denoted by R_p and it is smaller than the actual distance R travelled by ions. The relationship between R and R_p is approximated as

$$R_p \approx \frac{R}{1+\frac{m_2}{3m_1}} \tag{3}$$

where m_1 and m_2 are the masses of incident and target atoms, respectively. Lateral range is described as the distance travelled by the ions along the normal to the incidence direction and is denoted by $R^{\perp}$, while the spread in R caused by energy loss mechanism fluctuations is known as straggling. Ions stop at different depths in the target, and hence to find the distribution around R_p standard deviation or straggling (usually denoted by ΔR_p) is used [37]. The Lindhard, Scharff and Schiott (LSS) theory [38] describes the range and distribution of implanted ions into the target. According to the LSS theory, the distribution of implanted ions follows a Gaussian function. The concentration of implanted ions at a certain depth x can be expressed as

$$N(x) = N_{max}\, e^{-\frac{(x-R_p)^2}{2VR_p^2}}, \tag{4}$$

where N_{max} is the peak concentration of implanted ions and is given by

$$N_{max} = \frac{\Phi}{\sqrt{2\pi}\,V R_p} \approx \frac{0.4\Phi}{V R_p}, \tag{5}$$

where Φ is the ion dose in ion cm^{-2}, using this value of N_{max} we can find $N(x)$ as

$$N(x) = \frac{\Phi}{\sqrt{2\pi}\,V R_p}\, e^{-\frac{(x-R_p)^2}{2VR_p^2}} \tag{6}$$

Computer simulation is an accurate and fast method of finding the range, distribution and damage caused by implanted ions into a target. Transport and range of ions in matter (TRIM) is an extensive Monte Carlo binary collision computer simulation code based on full quantum mechanical treatment of ion-atom collisions [37]. TRIM can provide ion distribution together with all kinetic occurrences associated with the ion's energy loss (target damage, sputtering, ionization and phonon production) in multi-layered complex target materials.

Although Gaussian distribution is a logical estimation for the depth profile of dopants, it applies almost entirely for amorphous targets. Semiconductor crystals are highly crystalline, therefore, incident ions, if and when implanted parallel to crystal axes, move through crystal planes without experiencing any collision and are extremely likely to channel into the substrate. Therefore, to prevent channelling effects during implantation, the target crystal is tilted 7° of the direction of incident beam.

2.1. Thermal annealing

The implantation of energetic ions into semiconductors produces defects, damages the lattice and creates local amorphous regions and disordered crystalline structures. Notwithstanding several benefits, ion implantation can damage the lattice, which can cause quenching of luminescence, band filling, narrowing of band gap and band tailing effects. These effects acutely alter electrical and optical properties of devices. It is therefore necessary to anneal the samples after implantation to recover the lattice and at the same time move the implanted ions to suitable locations for electrical, optical and magnetic activation of the dopants. The implanted samples can be heated through rapid thermal annealing (RTA) up to 1000°C or even higher for a number of seconds in a controllable way. RTA offers very quick heating rates (1000°C per minute) and short time processing and this is its leverage over conventional furnace annealing (which takes several minutes or hours). Moreover, RTA provides relatively high security, taking into account sample decomposition, since the sample is exposed to high temperatures for a very short duration. High temperature annealing of III-nitrides is mostly carried out in a nitrogen atmosphere to prevent loss of nitrogen from the surface of the samples. RTA has been found as an efficient method to improve crystal quality and repair implantation-induced damage in III-nitrides.

The three major components of an RTA processor are RTA chamber, a heating system and a temperature sensor. RTA was carried out in an RTP-300 rapid thermal processor with 13 tungsten halogen lamps (1250 W each) as the source of heat and ambient nitrogen gas flow. A K-type thermocouple embedded in the sample stage was used as a temperature sensor. The samples to be annealed were placed upside down on another GaN wafer placed on a 4-inch silicon wafer positioned on a sliding sample holder inside the RTA chamber. After placing the sample holder inside the chamber, the window was tightly closed using screws and then RTA programmed for the required temperature and duration. Semiconductors are usually annealed up to a temperature as high as 2/3 of the melting point of bulk material. Hence, an annealing temperature of about 1800°C would be required for GaN whose melting point is 2791°C. However, 1800°C is higher than the growth temperature of epitaxial semiconductors and this would make the material decompose and deteriorate. Therefore, annealing temperatures and time need careful optimization to balance lattice recovery and sample degradation. A number of trial runs were performed using different annealing conditions in the search for optimized values of time and temperature. Here, a maximum annealing temperature of 900°C was used for a duration of 2–5 min.

2.2. Analysis techniques

2.2.1. X-ray diffraction

XRD is a very valuable technique to analyse the structure of crystalline materials. It provides an effective means of identifying crystal structures and investigates lattice modifications in the implanted/annealed samples. The X-ray diffractometer uses X-rays produced from the material due to shell-shell transitions as probes for analysis. These rays are produced when high energy electrons bombard on a copper target and give out a monochromatic beam of Cu-K_α radiation. When X-rays hit a crystalline material they are diffracted by the planes of the crystal. From Bragg's law, a diffraction peak is obtained only when the distance travelled by the rays after reflection from successive crystal planes differs by an integral multiple of wavelengths, in accordance with Bragg's equation:

$$2d\sin\theta = n\lambda \tag{7}$$

where d is inter-planer spacing, θ is the incident angle, λ is the wavelength of incident X-rays and n is the order of diffraction. The principle schematic diagram of an X-ray diffractometer is shown in **Figure 2**. A strong reflection or XRD peak is obtained by changing the angle θ so that the Bragg conditions are satisfied. Variation of angular positions with intensities of diffracted peaks produces a pattern peculiar to the material. Peak positions recorded in an XRD spectrogram are correlated with the peaks of known materials for phase analysis of the samples.

XRD analysis of the samples was performed using a Cu-K_α source of X-rays at room temperature by a Philips X'Pert data collector X-ray diffractometer. The crystallinity of the GaN samples was investigated in detail by carrying out $\omega/2\theta$ scans using double and triple axes diffraction. Moreover, peak broadening, tilt and twist characteristics were investigated by measuring FWHM from ω-scans of high resolution XRD. Powder diffraction XRD was carried out in 2θ ranges of 20–80° for phase analysis and detection of secondary phases in implanted samples.

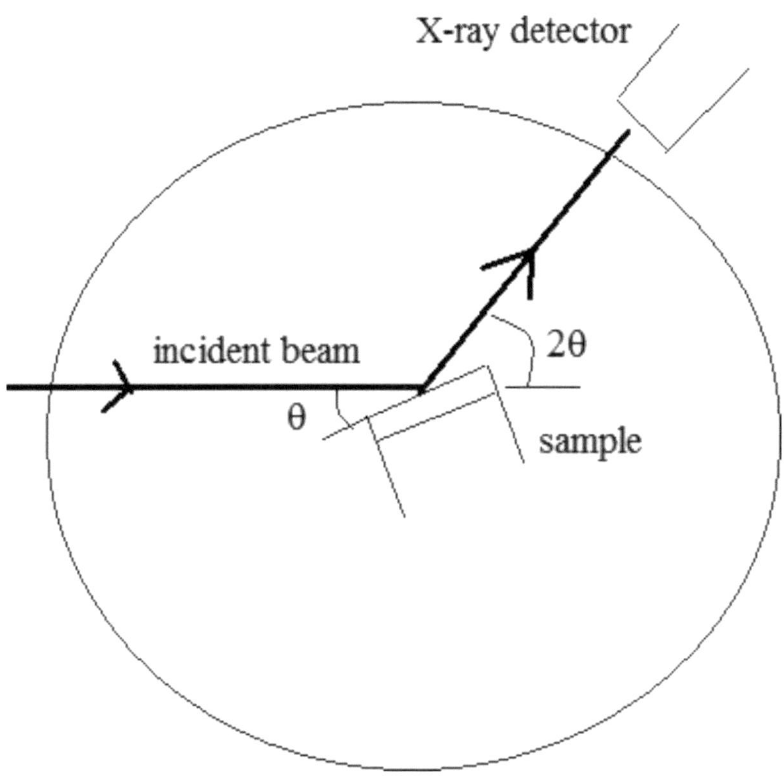

Figure 2. Principle and schematic diagram of diffractometer.

2.2.2. *High-resolution X-ray diffraction (HR-XRD)*

High resolution X-ray diffraction (HR-XRD) has long been used in the compound semiconductor industry for the characterization of epitaxial layers. The schematic diagram of HR-XRD is given in **Figure 3**.

Conventionally, HR-XRD has been employed to determine thickness and composition of epilayers, but of late the technique has progressed to enable the determination of strain and relaxation within a given layer on a multilayer structure. Typical HR-XRD symmetric reflections from a single layer on a bulk substrate are presented next.

A scan is taken by scanning sample and detector in 1:2 ratios. The substrate peak is normally the sharpest and most intense feature in the scan, also shown in **Figure 4**. The position of the Bragg peak is determined from Bragg's law. In this example, (**Figure 4**), a layer peak can be observed on the left-hand side of the substrate peak. This means that the lattice parameter of the layer is larger than that of the substrate since, from Bragg's law, it diffracts at smaller angles than the substrate. The differences in the positions of the peaks are related to the differences in lattice parameters, which can be due to composition, strain or relaxation of the layer. On both sides of the layer peak, there are interference fringes resulting from interferences of wave-fields in the layer. This information can be used for the accurate determination of the thickness of the layer.

HR-XRD measurements were performed at Beijing synchrotron radiation facility (BSRF) and Shanghai synchrotron radiation facility (SSRF). Synchrotron radiation is produced through the interaction of fast electrons with an applied magnetic field. The applied field will cause the electrons to accelerate by exerting a force on them perpendicular to their direction of motion, as shown in **Figure 5**. This will then cause the electrons to radiate electromagnetic energy

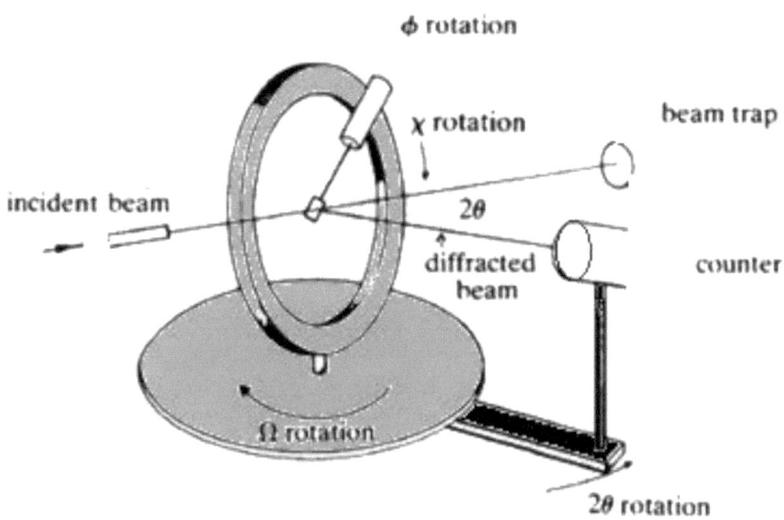

Figure 3. Principle and schematic diagram of HR-XRD.

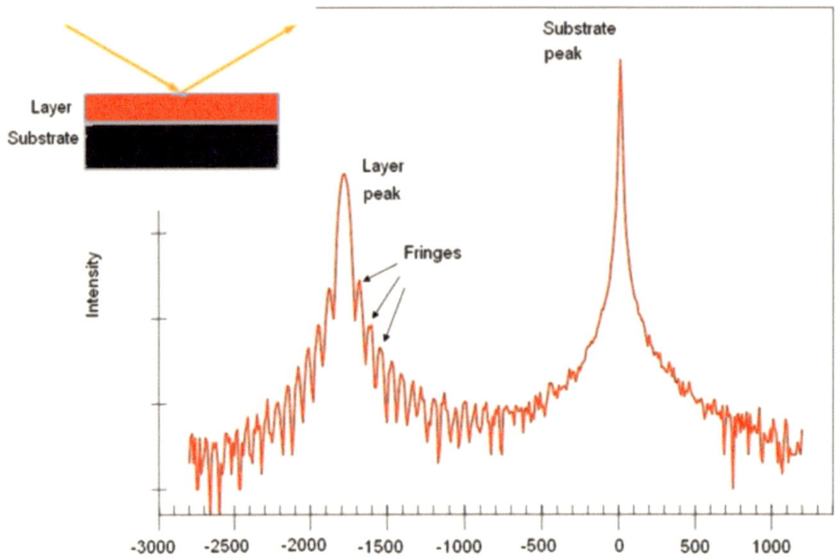

Figure 4. HR-XRD reflections from a single layer on a bulk substrate.

called magnetic bremsstrahlung or synchrotron radiation. If the energy of electrons and the magnetic field are high enough, X-rays can be produced.

2.2.3. Rutherford backscattering spectrometry (RBS)

Rutherford backscattering spectrometry (RBS) is an extensively used nuclear technique for near surface layer analysis of materials. Ions with energy in the MeV range (typically 0.5–4 MeV) bom-

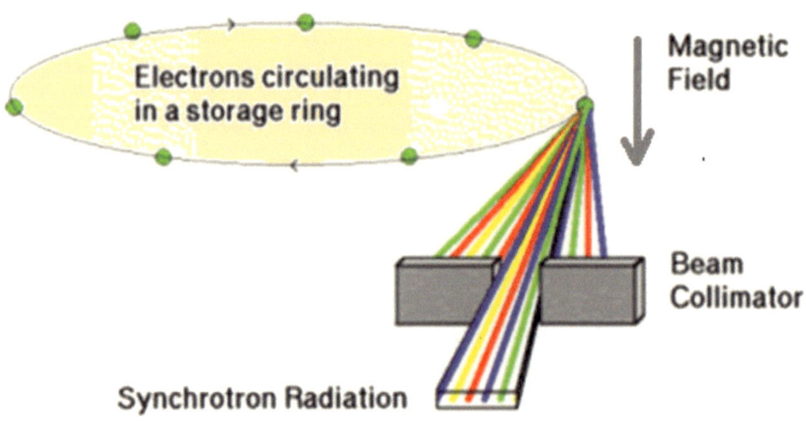

Figure 5. Principle of synchrotron radiation.

bard a target and backscatters. The energy of the backscattered ions is recorded by an energy sensitive detector, usually a solid-state detector. RBS permits the quantitative determination of material composition and a depth profile of individual elements is possible. It is quantitative without using reference samples, non-destructive and its depth resolution is good. The analysed depth is about 2 and 20 μm for He-ions and protons, respectively. The drawback of RBS is that its sensitivity for light elements is low, and this frequently requires complementing with other nuclear techniques such as nuclear reaction analysis (NRA) or elastic recoil detection analysis (ERDA).

The RBS technique is named after Sir Ernest Rutherford who, in 1911, used the backscattering of alpha particles from a gold foil to determine the fine structure of the atom, and this resulted in the discovery of the atomic nucleus. However, RBS as a materials analysis technique was first described in 1957 by Rubin et al. [39]. The book by Tesmer et al. [40] is highly recommended for further reading on modern applications of the RBS technique. RBS encompasses all forms of elastic ion scattering at incident ion energies ranging from about 500 keV to several MeV. Normally, protons, alpha particles and sometimes lithium ions are used as projectiles at backscattering angles of between 150 and 170°. There are special cases where different angles or projectiles are used. When inelastic scattering and nuclear reactions are used, the method is called nuclear reaction analysis (NRA), while detection of recoils at forward angles is called elastic recoil detection analysis (ERD or ERDA). Due to the long history of RBS, there have been many and sometimes uncontrolled growth of acronyms and a list of recommended ones can be found in Amsel [41].

2.2.3.1. *Scattering geometry and kinematics*

Figure 6 shows the most commonly used scattering geometries. If the incoming beam, outgoing beam and the surface normal to the sample are in the same plane, then we get the IBM geometry. And the relationship between the incident angle α, exit angle β and scattering angle θ is given by

$$\alpha + \beta + \theta = 180^{\circ}, \tag{8}$$

while in the Cornell configuration, the incoming beam, outgoing beam and the sample rotation axis are in the same plane, and

$$\cos(\beta) = -\cos(\alpha)\cos(\theta). \tag{9}$$

The advantage of the Cornell geometry is that it combines a large angle of scattering, which is advisable for better mass resolution, and grazing incident and exit angles, which improves depth resolution. If a projectile with incident energy E_0 and mass M_1 backscatters from a target, its energy E_1 after scattering is given in the laboratory system by

$$E_1 = K E_0, \tag{10}$$

where the kinematic factor K is given by

$$K = \frac{M_1^2}{(M_1 + M_2)^2} \left\{ \cos\theta \pm \left[\left(\frac{M_2}{M_1} \right)^2 - \sin^2\theta \right]^{1/2} \right\}^2 \tag{11}$$

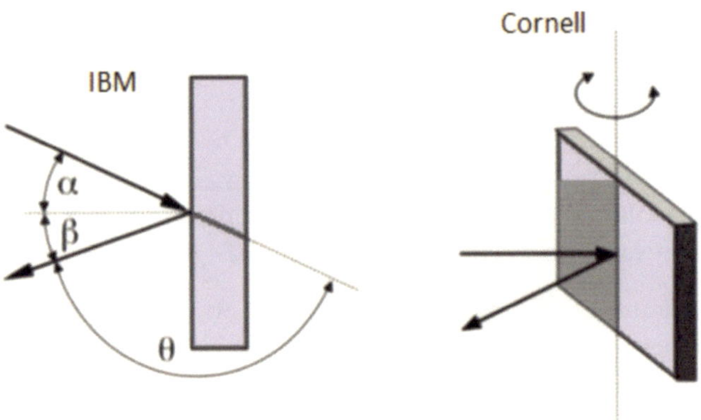

Figure 6. Left: IBM geometry; right: Cornell geometry. Incident angle α, exit angle β and scattering angle θ.

where θ is the scattering angle and M_2 is the mass of the target nucleus, initially at rest. The plus sign in above equation applies when $M_1 < M_2$. If $M_1 > M_2$ then the equation has two solutions, and the maximum possible scattering angle θ_{max} is given by

$$\theta_{max} = \arcsin\left(\frac{M_2}{M_1}\right) \tag{12}$$

Figure 7 shows the relationship between the kinematic factor and target mass for various incident ion beams.

The RBS technique uses the Rutherford scattering principle. If a high energy helium ion beam ($^4He^+$) is backscattered from a target, then the collision probability of the incident alpha particles with atoms of the target is determined by the Rutherford cross-section. If there is elastic

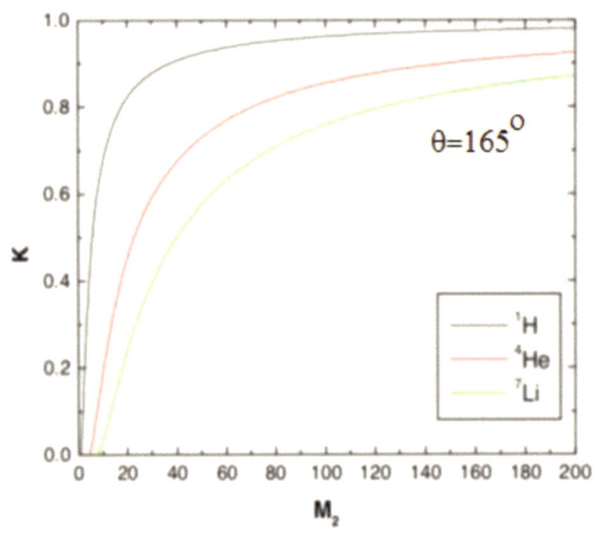

Figure 7. Kinematic factor K at a scattering angle θ = 165° as a function of target mass $M2$.

collision between incident alpha particles and target atoms then the energy ratio of the particles pre- and post-collision is equal to the ratio of the masses of both particles. Information on the atoms from which the alpha particles have been backscattered can be inferred from the energy of the backscattered alpha particles. According to the single scattering theory, an alpha particle faces only one large angle scattering before reaching the detector. The principle and schematic diagram of the RBS technique is shown in **Figure 8**. This approximation helps to convert the energy scale into the depth scale of the sample within the energy resolution of the detector. The lower the energy of the backscattered alpha particles, the deeper the detected atoms.

In this study, an alpha beam of energy 2 MeV with a diameter of 1 mm was used. The samples were mounted on a two axis goniometer, which can align the sample with the incident beam at any angle required. There are two silicon detectors in the target chamber. One detector is used to measure particles backscattered at angles near the incident beam and the second detector is for particles backscattered at glancing angles to the sample surface. The second detector, set at 165° relative to incident beam direction and 80 mm away from the sample, has a resolution of 18 keV and an aperture diameter of 5 mm. The RBS data was analysed using RUMP simulation.

2.2.4. *AGM and SQUID*

To manipulate magnetic materials, it is very important to know their magnetic moments and this can be measured with the alternating gradient magnetometer (AGM). Although AGM is extremely sensitive, it cannot measure single magnetic markers directly. Alternatively, the average magnetic moment for a single bead is calculated from measurements of several millions of markers. Additionally, the number of measured magnetic markers cannot be counted exactly, but only estimated by the given dilution. The magnetic moment at a small outer field (~100 Oe), not the moment for saturated magnetic beads, is interesting for bond-force measurements. Apart from that only the mean magnetic moment of the beads should be known, more problems were found during measurements. Although the beads are superparamagnetic, some of them show remanent magnetization. This could be due to clustering of the beads, not fully oxidized magnetite (Fe_3O_4) particles inside the beads or a few very big beads. In order to prevent clustering, the magnetic markers are pipette spotted onto a heated Si-wafer (~100). Clustering could not have been the sole reason for ferromagnetic behaviour since the effect remained. Furthermore, the magnetic moment

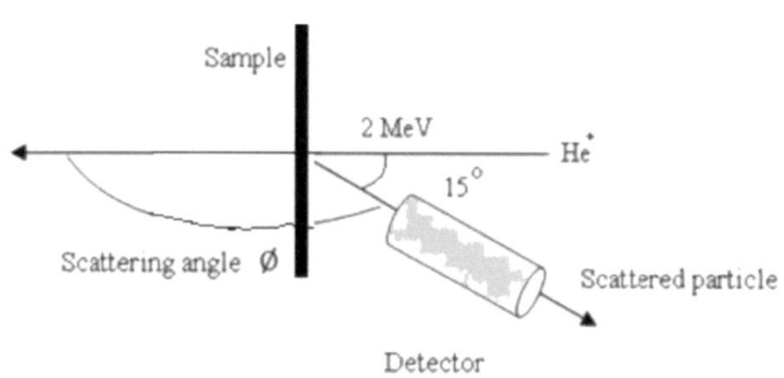

Figure 8. Principle and schematic diagram of RBS.

shows dependency on bead concentrations, which cannot be neglected (six different concentrations were tested for every bead type). All of this has to be considered when calculating the magnetic moment at a small outer magnetic field.

SQUID magnetometry is a very sensitive technique for magnetic characterization of materials. A SQUID can detect magnetic ordering by tracking temperature dependency on magnetization (MT), field-dependent magnetization (MH) and very weak magnetic moments. A SQUID works on the principle of electron-pair wave coherence and Josephson effect, which can be defined as the flow of current (called Josephson's current) across two superconductors separated by an insulated layer. A Josephson's junction comprises two superconducting coils separated by a very thin insulating barrier to enable electrons to pass through it. A SQUID magnetometer is made of a superconducting ring into which two Josephson's junctions are placed in parallel in a magnetic field. A current flows through the superconducting loop if a magnetic field is applied. The magnetic flux of a ferromagnetic sample placed between the superconductors in the presence of an applied field will change accordingly. This magnetic flux change will induce a current which changes the initial current circulating through the coil. The variation in the current helps to detect the magnetic moment of the material.

Figure 9 shows the principle of SQUID magnetometry. A SQUID magnetometer was used to investigate MT and MH characteristics of metal ion-implanted GaN samples. During MH analysis, the hysteresis loops of ion-implanted and unimplanted samples were recorded at

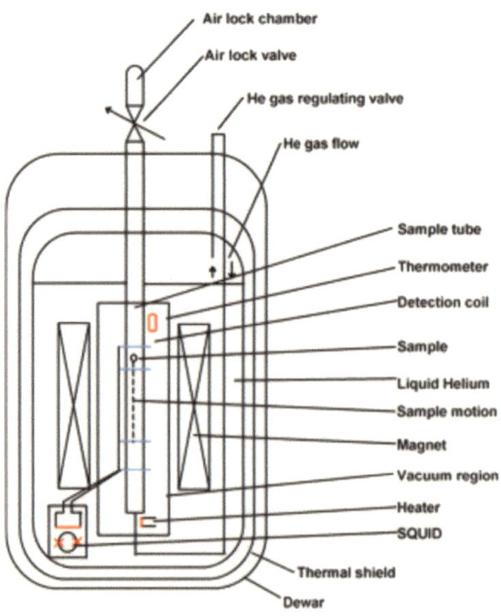

Figure 9. Schematic diagram of a typical SQUID magnetometer [42].

100 and 300 K, respectively. For MT measurements, the sample was first cooled down to 5 K with no applied magnetic field. A 500 Oe magnetic field was then applied and a scan collected up to 350 K to obtain a zero field cooled (ZFC) trace. Secondly, the sample was cooled down to 5 K in the 500 Oe field and magnetization measured to get the field cooled (FC) trace. The surface of the sample was kept parallel to the applied magnetic field during magnetization measurements.

3. Methodology and materials

Epitaxial layers of GaN (n-type) about 2 μm thick were grown on sapphire (Al_2O_3) by metal organic chemical vapour deposition (MOCVD). The structure and crystal quality of the GaN epilayers were analysed by Rutherford backscattering and channelling spectrometry (RBS/C) prior to ion implantation. A He^+ ion beam, with energy 2.0 MeV, was used for RBS/C analysis. Sample mounting and detector specifications and configurations have already been explained in Section 2.2. The <0001> normal axis was chosen to investigate the quality of GaN. The channelling minimum yield was found to be around 1.3%, indicating that the quality of the epitaxial GaN material was high. Co^+ ions of energy 150 keV were implanted into the GaN epilayers at room temperatures at doses of 3×10^{16} and 5×10^{16} ions cm^{-2}, making the wafer to self-heat at such high doses. This is advantageous since it causes 'dynamic annealing', that is, annealing during implantation. The concentration of cobalt was calculated to be around 3–5% from SRIM 2008 [37]. The estimated range of Co^+ ions in the sample was approximately 150 nm. The implanted samples were annealed at 700, 800 and 900°C for 5 min using rapid thermal annealing (RTA) in an ambient N_2 atmosphere to remove implantation damage and recrystallize them. Similarly, Cr^+ ions of energy 150 keV were implanted into the GaN epilayer at room temperature to a dose of 3×10^{16} ions cm^{-2} at a tilted angle of 7° of the direction of incident beam to avoid implantation channelling effects. The chromium concentration was estimated to be about 3% as calculated from SRIM 2008 and also confirmed by simulation of random spectra using RUMP. The projected range of the Cr^+ ions in the samples was about 150 nm. The implanted samples were annealed at 800 and 900°C for 2 min using rapid thermal annealing (RTA) in ambient N_2 to re-crystallize the samples and to remove implantation damage.

X-ray diffraction (XRD, Philips X'Pert Data Collector) was used for structural analysis of ion-implanted samples, using Cu K_α radiations. High resolution X-ray diffraction (HR-XRD, BEPCII-2.5 GeV) was performed at Beijing synchrotron radiation facility (BSRF) and at Shanghai synchrotron radiation facility (SSRF). Alternating gradient magnetometry (AGM, PMC 2900-04C) was used for measuring magnetization at room temperature and a superconducting quantum interference device magnetometer (MPMS-XL Quantum Design) was used to investigate magnetic properties at 5 K. A magnetic field was applied parallel to the sample surface during magnetization measurements.

4. Metal ion implantation into GaN

Cobalt ions were implanted into GaN at doses of 3×10^{16} and 5×10^{16} ions cm^{-2} and chromium ions were implanted at a dose of 3×10^{16} ions cm^{-2}.

4.1. Cobalt implantation into GaN

4.1.1. X-ray diffraction (XRD)

Typical XRD spectra of GaN for as-grown and all the implanted samples annealed at 700, 800 and 900°C are given in **Figures 10** and **11**. **Figures 10** and **11** show typical XRD profiles of Co^+ implanted GaN at 3×10^{16} and 5×10^{16} ions cm^{-2} and subsequently annealed at 700, 800 and 900°C. In the as-grown sample, three main peaks appeared corresponding to the expected diffraction from the GaN epilayer and sapphire substrate structure.

Comparison of the XRD patterns of the as-grown with the implanted samples at different doses, it can be observed that no secondary phases or metal-related peaks were detected in the as-implanted samples and annealed samples. The diffraction patterns show peaks corresponding to the GaN layer and the substrate structure only. However, the presence of sufficiently small cobalt nanoscale precipitates, which cannot be detected by XRD due to its insensitivity on the nanoscale [43, 44], is possible.

HR-XRD ($\theta - 2\theta$) spectra showing the (0002) peak of GaN for the as-grown and selected implanted samples at doses 3×10^{16} and 5×10^{16} ions cm^{-2} and annealed at 900°C are given in **Figures 12** and **13**. In **Figure 12**, the diffraction pattern of the implanted sample, a typical satellite peak appears at lower side of the main GaN (0002) reflection. Ion implantation into crystalline GaN introduces lattice disorder which is a side effect of implantation [4]. As a result, in addition to GaN peak, new peak/peaks, representing the damaged part of lattice, appear on the low angle side of the main GaN peak in the XRD spectra of implanted GaN as reported by many researchers [45–49].

The shape, position and number of such new peaks were found to differ for different ions implanted into GaN. Most of the authors attributed such new XRD peaks to the implantation induced strain and the expansion of GaN lattice in the implanted portion of the material [45–48]. Another group of researchers suggested that these peaks were related to the formation of new phases [49]. Liu et al. presented a comparative XRD study of Ca- and Ar-implanted GaN and observed larger lattice expansion for Ar implantation. They assigned the observed phenomenon to the inability of inert gas ions to occupy substitutional sites in the lattce [45]. Inert gases, due to their very low solubility in solids, are reported to produce small gas-vacancy clusters that lead to the formation of gas-filled cavities called bubbles [50]. The formation of such inert gas cavities was also observed in several other materials such as Si [51], GaAs [52], SiC [53] and InP [54]. These empty cavities, due to their negative curvatures, contain high density of dangling bonds that exhibit high affinity for metallic contaminants and can act as impurity gettering sites [55]. Gettering of oxygen impurity atoms and structural defects in GaN by helium implantation has been reported [55–57].

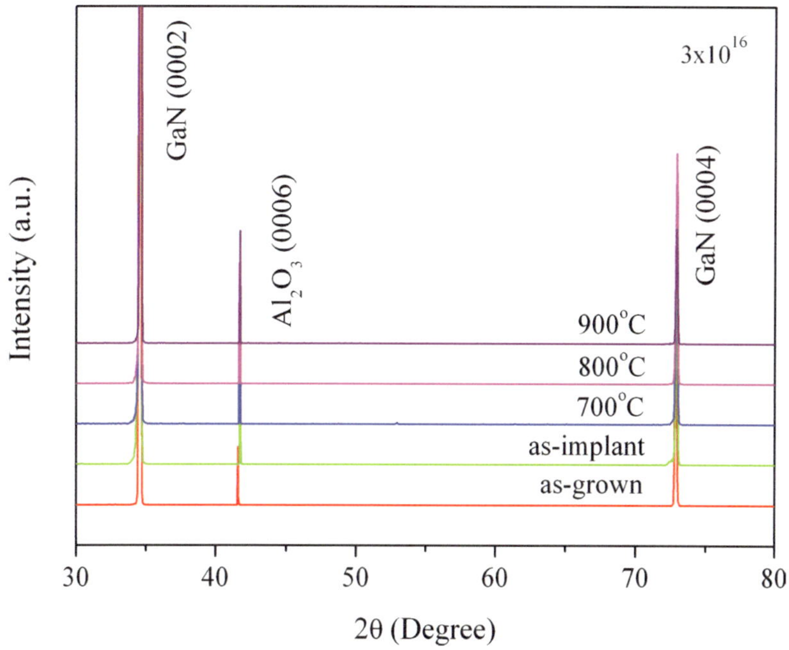

Figure 10. Typical XRD pattern of Co+ implanted at dose 3 × 10^{16} cm^{-2} [66].

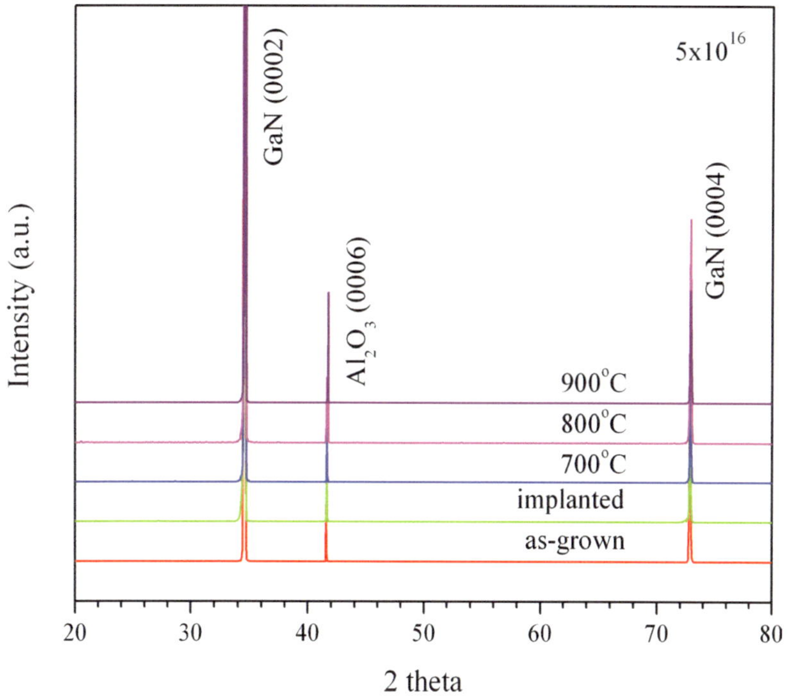

Figure 11. Typical XRD pattern of Co+ implanted at dose 5 × 10^{16} cm^{-2} [67].

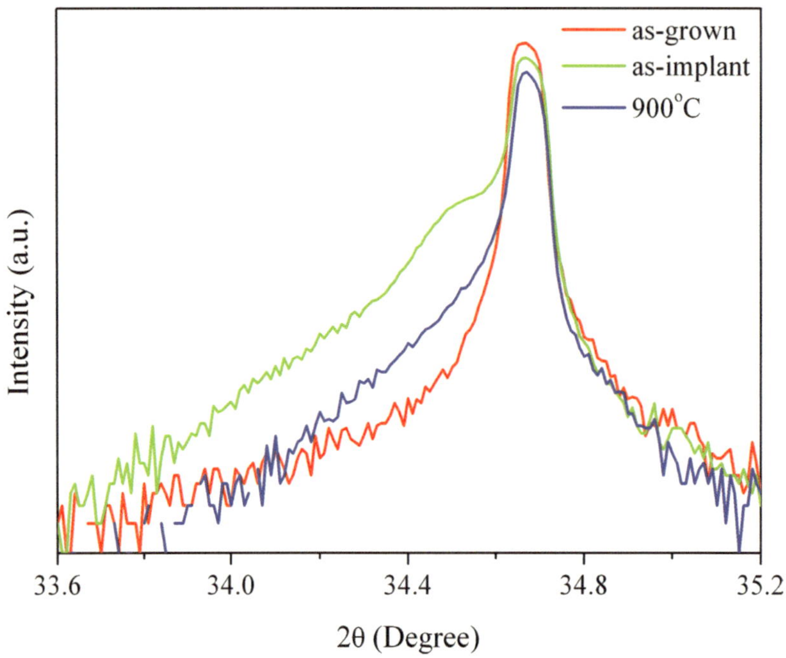

Figure 12. HR-XRD pattern of Co+ implanted at 3 × 10^{16} ions cm^{-2} [66].

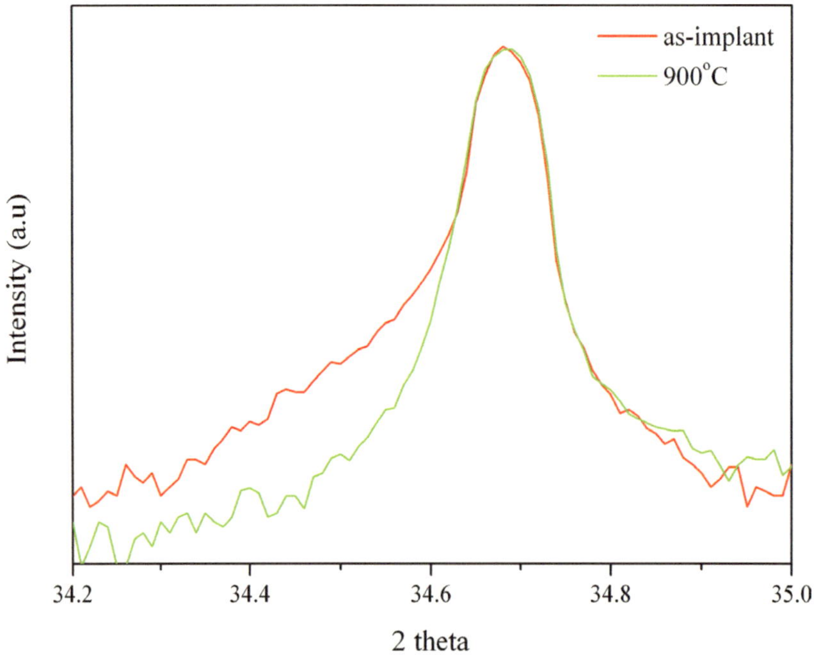

Figure 13. HR-XRD pattern of Co+ implanted at 5 × 10^{16} ions cm^{-2} [67].

http://dx.doi.org/10.5772/68042

The presence of similar peaks has previously been observed in the XRD spectra of GaN implanted with different ions and is thought to be due to lattice expansion along the *c*-axis of GaN [45, 46]. The new peak from HR-XRD measurements is attributed to implantation-induced damage and also to the formation of $Ga_{1-x}Co_xN$ on the part of the sample which was implanted. A shoulder peak observed on the left side of the GaN peak on XRD scans of MBE grown samples has been attributed to the GaMnN phase by Cui and Li [58]. The lattice constant of (Ga,Co)N varies with cobalt incorporation, implying that the position of the new peak is related to the amount of cobalt in the material. Hence, the introduction of cobalt at interstitial and substitutional sites in GaN is expected to cause lattice expansion to produce a new XRD peak on the left side of the GaN peak [59]. The shifting of additional peaks to the right with annealing, presented in both **Figures 12** and **13**, points to lattice recovery and improvement in the uniformity of GaCoN which may be due to increase in the substitution probability of cobalt atoms.

4.1.2. RBS channelling

The RBS channelling spectrum of the as-grown GaN/sapphire (0001) together with the corresponding minimum channelling yield χ_{min} = 1.3%, indicating high crystalline quality [60, 61], is shown in **Figure 14**.

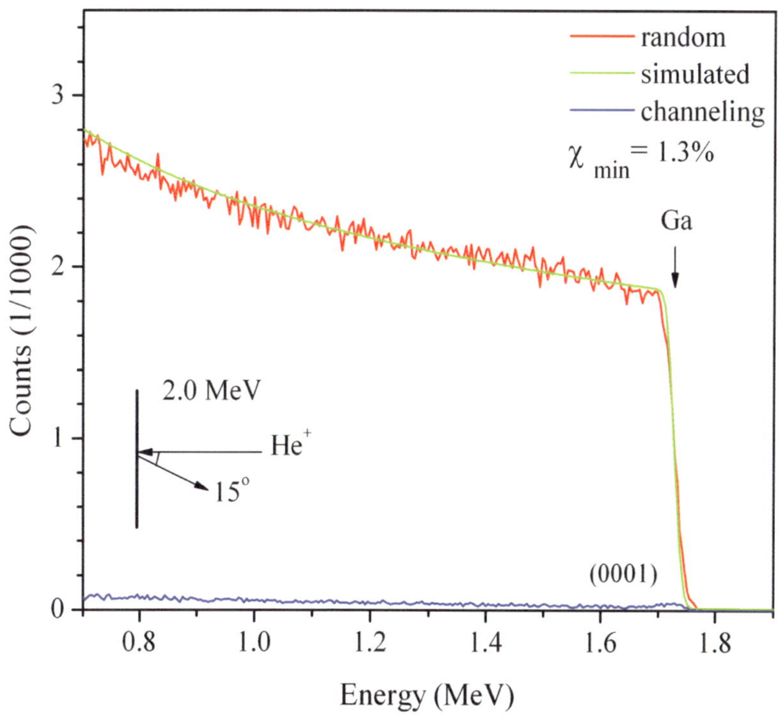

Figure 14. RBS/C of as-grown sample along with the backscattering geometry [66].

A random spectrum simulation was carried out using the RUMP program [62]. Channelling spectra are presented in **Figures 15** and **16** for Co^+ implanted GaN sample at doses of 3 × 10^{16} and 5 × 10^{16} ions cm^{-2} and subsequently annealed samples at 700, 800 and 900°C along with the corresponding minimum channelling yields χ_{min}. **Figures 15** and **16** present minimum channelling yields χ_{min} calculated for the maximum at around 1.64 MeV and is related to the random spectrum of virgin (upper spectrum) GaN.

The random spectrum of the as-implanted GaN is not shown here due to minor differences with the random spectrum of as-grown GaN. Co^+ implanted GaN sample at doses 3 × 10^{16} and 5 × 10^{16} ions cm^{-2} and subsequently annealed at 900°C showed better recovery of implantation damage. From our measurements, annealing at 900°C is the most suitable annealing temperature to re-crystallize the samples.

4.1.3. AGM and SQUID

Magnetization against magnetic field (M-H) curves from AGM measurements at room temperature for the samples implanted at doses of 3 × 10^{16} and 5 × 10^{16} ions cm^{-2} and subsequently annealed at 700, 800 and 900°C are shown in **Figures 17** and **18**, where the signal from the sapphire substrate (diamagnetic) was extracted. The magnetic field was applied parallel to the sample plane. Well-defined hysteresis loops were observed even at 300 K. The saturation field was about 4000 Oe and the coercivity H_c was about 100 Oe for the implanted and annealed samples. These results confirm that the samples were ferromagnetic even at room temperature.

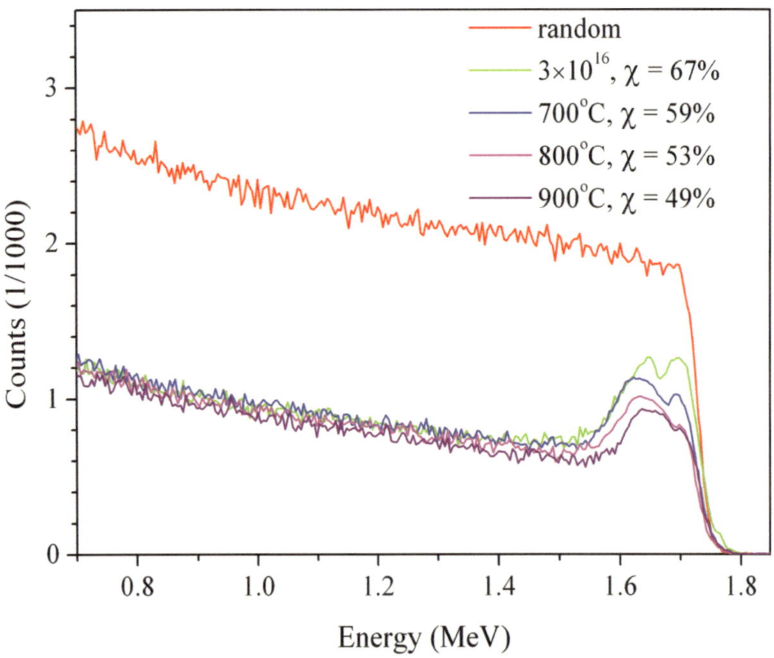

Figure 15. RBS/C of Co+ implanted GaN at 3 × 10^{16} ions cm^{-2} and annealed samples at different temperatures [66].

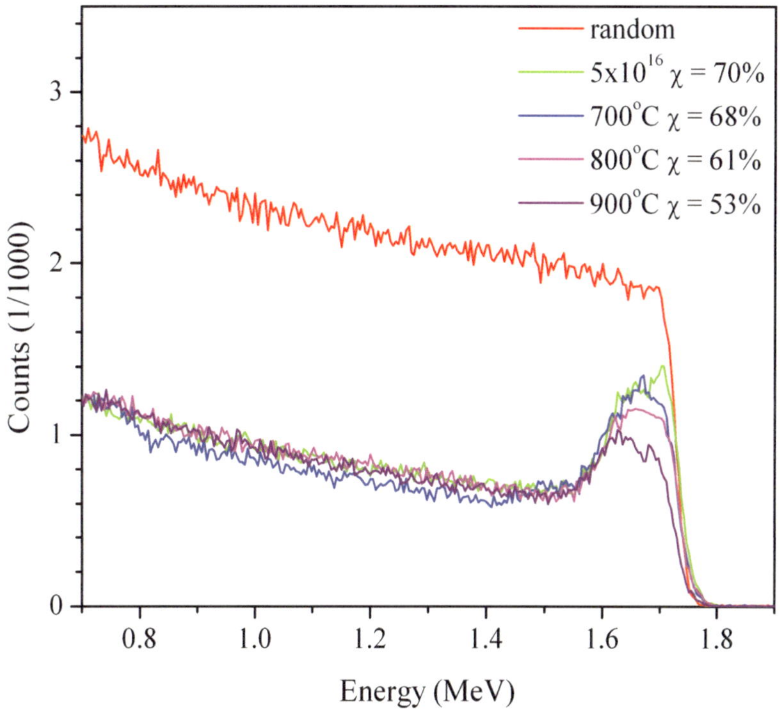

Figure 16. RBS/C of Co+ implanted GaN at 5 × 10^{16} ions cm^{-2} and annealed samples at different temperatures [67].

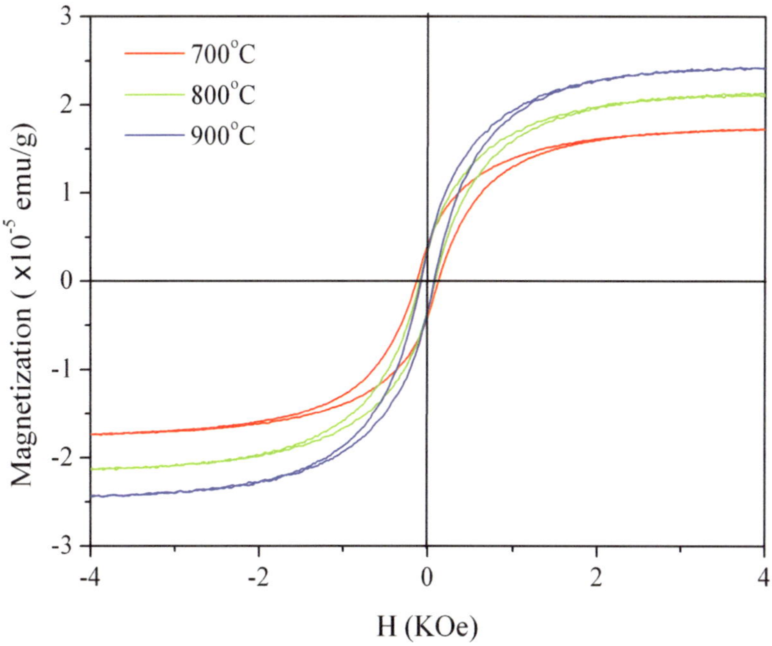

Figure 17. M-H loops at 300 K of sample implanted at dose 3 × 10^{16} ions cm^{-2} and annealed [66].

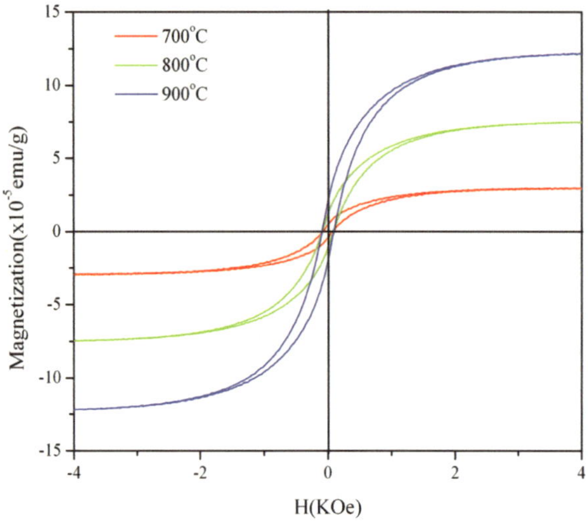

Figure 18. M-H loops at 300 K of sample implanted at dose 5 × 10^{16} ions cm^{-2} and annealed [67].

Similarly, a well-defined hysteresis loop, measured using a SQUID magnetometer, at 5 K was also observed from samples implanted at a dose of 3 × 10^{16} ions cm^{-2} and subsequently annealed at 700, 800 and 900°C as shown in **Figure 19**. The saturation field was about 4000 Oe and the coercivity H_c was about 180 Oe for implanted and annealed samples at 700 and 800°C and the coercivity was observed around 270 Oe for the sample annealed at 900°C. Also, a well-defined hysteresis loop at 5 K from SQUID was also observed for the samples implanted at a dose of 5 ×10^{16} ions cm^{-2} and subsequently annealed at 700, 800 and 900°C as shown in **Figure 20**. The saturation field was about 4000 Oe and the coercivity *Hc* was about 270 Oe for implanted and annealed samples at 700 and 800°C and the coercivity was observed around 600 Oe for the sample annealed at 900°C.

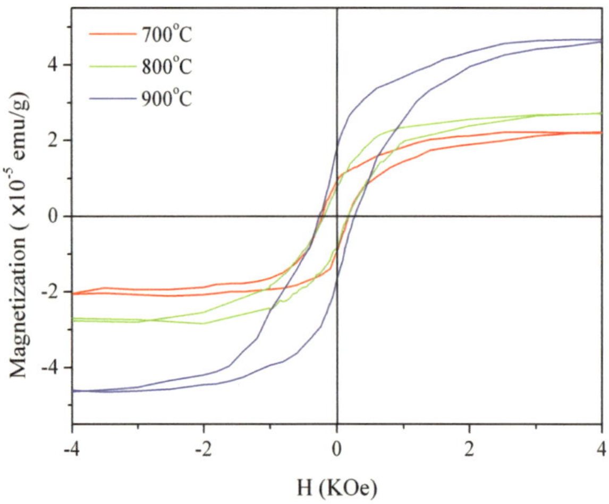

Figure 19. M-H loops at 5 K of sample implanted at dose 3 × 10^{16} ions cm^{-2} and annealed [66].

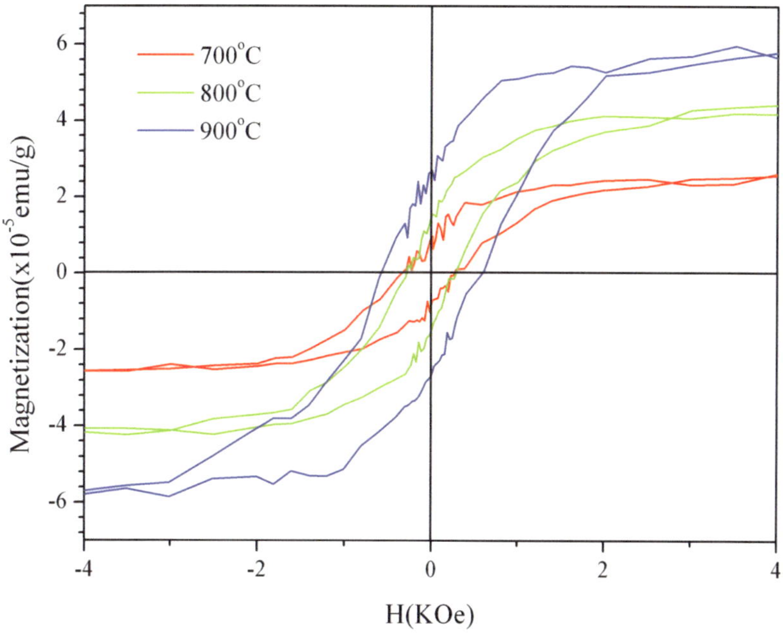

Figure 20. M-H loops at 5 K of sample implanted at dose 5 × 10^{16} ions cm^{-2} and annealed [67].

Magnetization as a function of temperature for selected samples is plotted in **Figures 21** and **22**. The variation of magnetization with temperature indicates multiple exchange interactions,

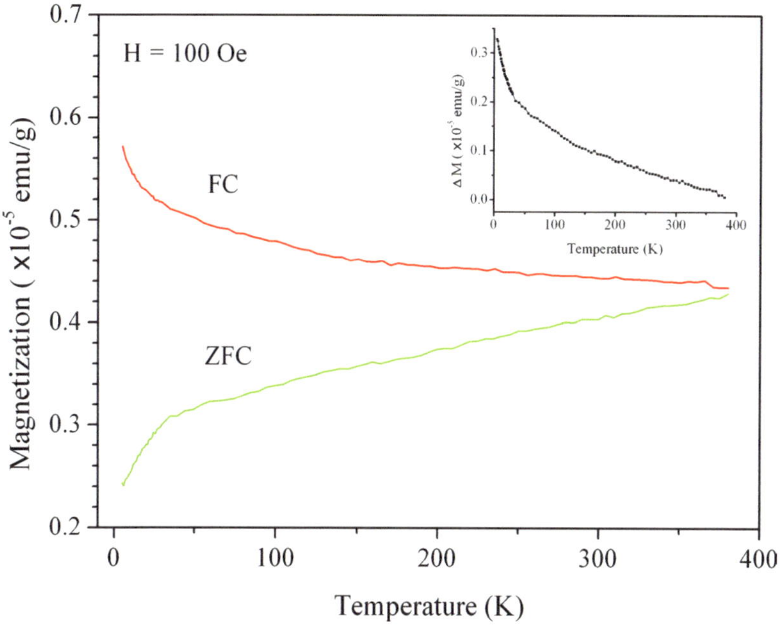

Figure 21. FC/ZFC measurements of sample implanted at a dose of 3 × 10^{16} ions cm^{-2} and annealed at 900°C [66].

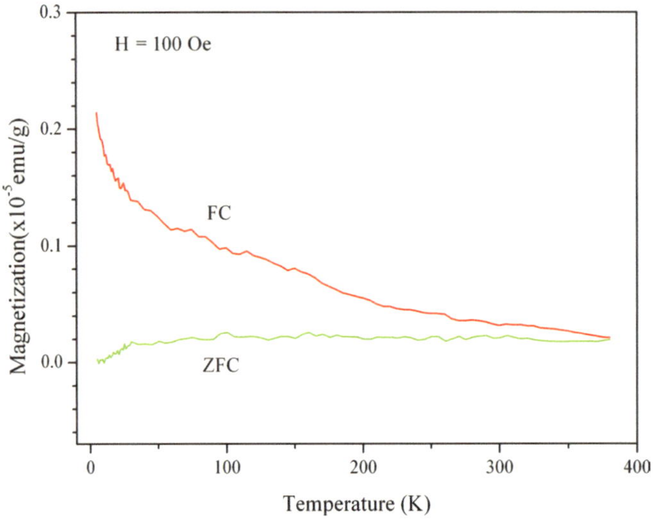

Figure 22. FC/ZFC measurements of sample implanted at a dose of 5 × 10^{16} ions cm^{-2} and annealed at 900°C [67].

indicating that its decay cannot be easily fit to classical description of ferromagnetism, again in agreement with current theories concerning DMS systems with low carrier concentrations. Co^+ implanted GaN at doses of 3 × 10^{16} and 5 × 10^{16} ions cm^{-2} and annealed at 900°C showed magnetic moment at lower temperatures and retained magnetization up to 370 K. There were indications of the possible presence of multiple complex exchange interactions for Co^+ implanted GaN. The same phenomenon has been observed on Cr^+ implanted GaN.

All samples exhibited well-saturated MH loops (**Figures 17–20**) with finite coercivity, eliminating the likelihood of both paramagnetism and superparamagnetism [63]. Analysing hysteresis loops of the implanted samples assists in the investigation of the magnetic properties of the material. Lateral shifting of hysteresis loop was not observed, and this eliminates spin glass behaviour [64]. These observations imply that there was ferromagnetic ordering in implanted samples at room temperature. No extra peaks were observed on the XRD spectra of the implanted samples (**Figures 10** and **11**), reducing the contributions of secondary phases (Co_xN_y, CoGa, etc.) to the observed ferromagnetism. FC and ZFC measurements were performed on a representative sample, together with magnetization as a function of temperature, and the data did not show any blocking temperature that can be related to superparamagnetic behaviour arising from undetected magnetic secondary phase clusters.

4.2. Chromium implantation into GaN

4.2.1. X-ray diffraction (XRD)

Typical XRD spectra of GaN for the as-grown and Cr^+ implanted at a dose of 3 ×10^{16} ions cm^{-2} and subsequently annealed at 800 and 900°C are given in **Figure 23**. In the as-grown sample, the three main peaks correspond to the expected diffraction from the GaN epilayer and sapphire substrate structure.

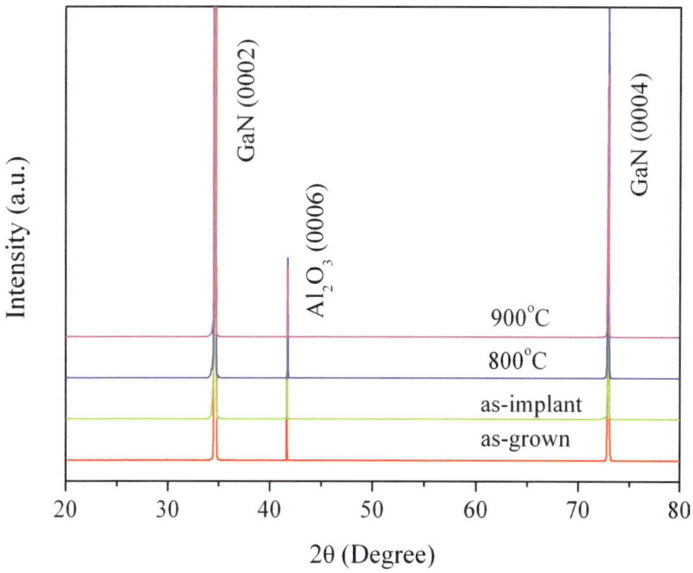

Figure 23. Typical XRD pattern for the as-grown, implanted and annealed samples at different temperatures.

XRD did not show any secondary phases or metal-related peaks on the as-implanted samples and annealed samples, when compared with the as-grown sample. Only peaks corresponding to the GaN layer and the substrate structure could be observed on the diffraction pattern. However, the presence of sufficiently small chromium nanoscale precipitates, which cannot be measured by typical XRD due to its insensitivity on the nanoscale [43, 44], is not excluded. HR-XRD spectra of GaN for implanted samples at a dose of 3×10^{16} ions cm^{-2} and annealed at 900°C are given in **Figure 24**. In the diffraction pattern of the implanted sample, a typical satellite peak appears at the lower side of the main GaN (0002) reflection.

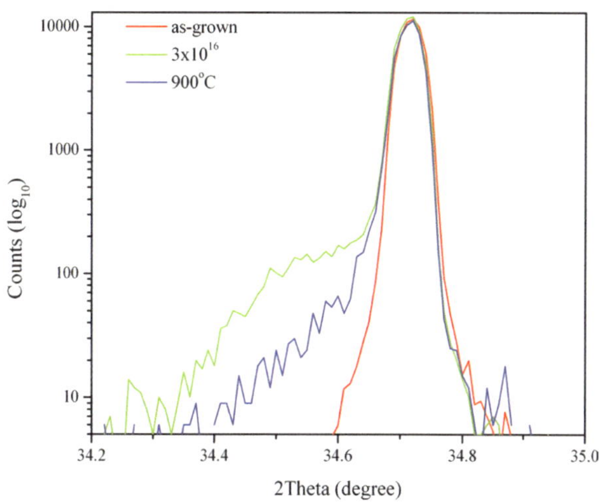

Figure 24. 2 HR-XRD spectra for Cr+ implanted GaN and subsequently annealed at 900°C [68].

The shape, position and number of such new peaks were found different as observed in Co^+ implanted GaN epilayers. The new peak in the XRD scans is assigned to implantation-induced damage as well as the formation of $Ga_{1-x}Cr_xN$ in the implanted part of the sample. The lattice constant of (Ga,Cr)N changes due the presence of chromium, implying that the position of the new peak is related to the concentration of chromium in the material. Hence, the introduction of chromium at interstitial and substitutional sites in GaN is expected to cause lattice expansion to produce a new XRD peak on the left side of the GaN peak [59]. The shifting of additional peaks to the right with annealing, presented in **Figure 24**, points to lattice recovery and improvement in the uniformity of GaCrN which may be due to increase in the substitution probability of chromium atoms.

4.2.2. RBS channelling

Channelling spectra are presented in **Figure 25** for Cr^+ implanted GaN sample at dose 3×10^{16} ions cm^{-2} and subsequently annealed at 800 and 900°C along with the corresponding minimum channelling yield χ_{min}. A minimum channelling yield χ_{min} was calculated for the maximum at around 1.65 MeV and is related to the random spectrum of the as-implanted sample (upper spectrum). Cr^+ implanted GaN sample at dose 3×10^{16} ions cm^{-2} and annealed at 800 and 900°C showed the recovery of implantation damage. If we compare these results with Co^+ implanted samples at same dose, we observe that there is less recovery of implantation damage by annealing using RTA for Cr^+ implanted samples. This may have been due to shorter annealing time (2 min).

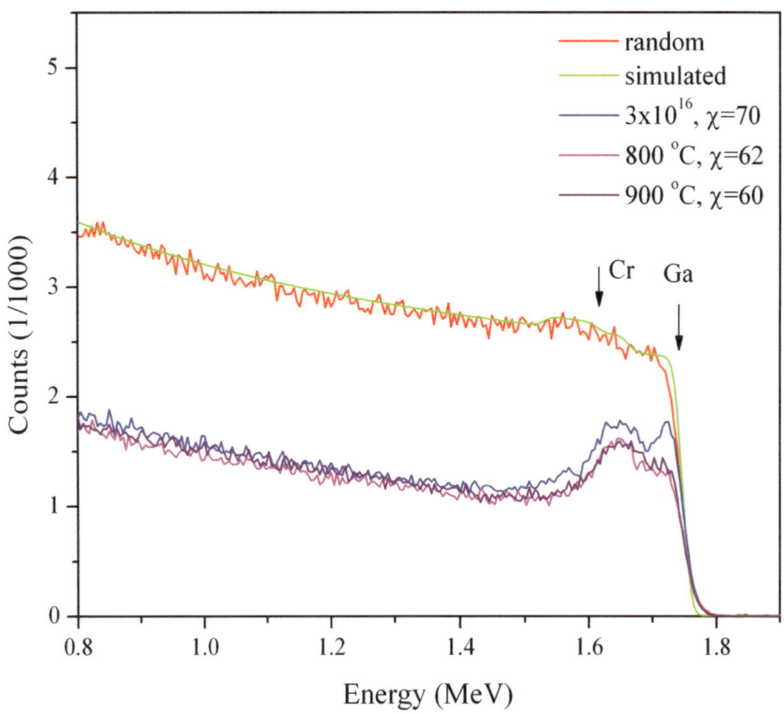

Figure 25. RBS/C of as-implanted sample at 3×10^{16} ions cm^{-2} and annealed at 800 and 900°C [68].

http://dx.doi.org/10.5772/68042

4.2.3. AGM and SQUID

Magnetization versus magnetic field (M-H) curves from AGM measurements at room temperature for the sample implanted to a dose of 3 $\times 10^{16}$ ions cm^{-2} and annealed at 800 and 900°C are shown in **Figure 26**, where the signal from the sapphire substrate (diamagnetic) was extracted. The magnetic field was applied parallel to the sample plane. At 300 K, a well-defined hysteresis loop was observed, which provides evidence for the presence of ferromagnetic interactions at room temperature. The saturation field was about 4000 Oe and the coercivity H_c was about 100 Oe for the implanted and subsequently annealed samples. These results confirm that the samples were ferromagnetic even at room temperature. Comparison with same dose for Co^+ implanted samples shows that the saturation magnetization *Ms* values are almost the same in all samples.

Similarly, a well-defined hysteresis loop at 5 K from SQUID was also observed for the implanted and subsequently annealed samples at 800 and 900°C, as shown in **Figure 27**. The saturation field is about 4000 Oe and the coercivity H_c is about 175 Oe for implanted and annealed samples. If we compare the results at 5 K with the same dose of Co^+ implanted samples we find that the saturation magnetization *Ms* values are almost similar for the sample annealed at 800°C. But the *Ms* value for Cr^+ implanted sample annealed at 900°C is 10.7 ($\times 10^{-5}$ $emug^{-1}$) while for the Co^+ implanted it is about 4.5 ($\times 10^{-5}$ $emug^{-1}$). A higher value of *Ms* for Cr^+ implanted GaN epilayer annealed at 900°C suggests that samples implanted with Cr^+ ions may perform better for dilute magnetic semiconductors (DMSs) compared to Co^+ implanted. Also higher values of *Ms* for implanted samples may suggest that 900°C is a suitable annealing temperature for the activation of dopants.

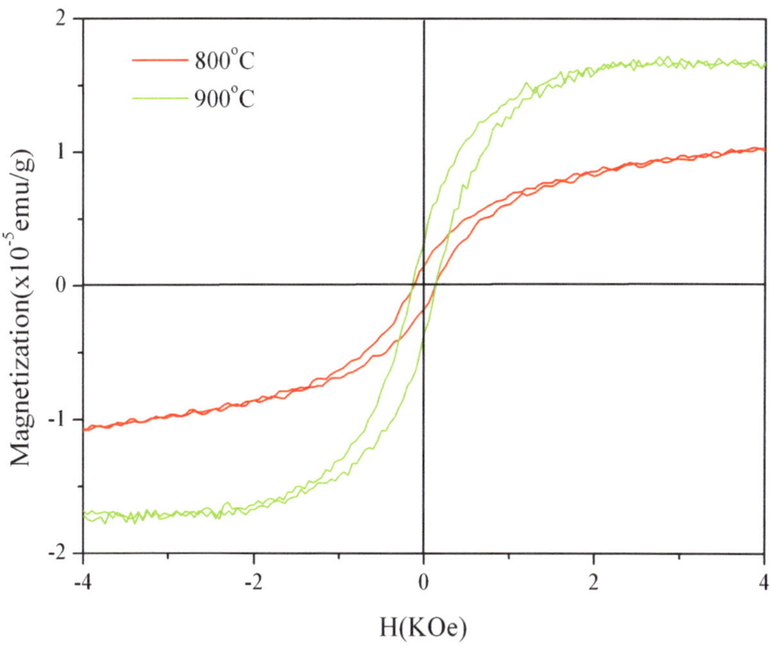

Figure 26. M-H loops at 300 K of sample implanted at dose 3 × 10^{16} ions cm^{-2} and annealed [68].

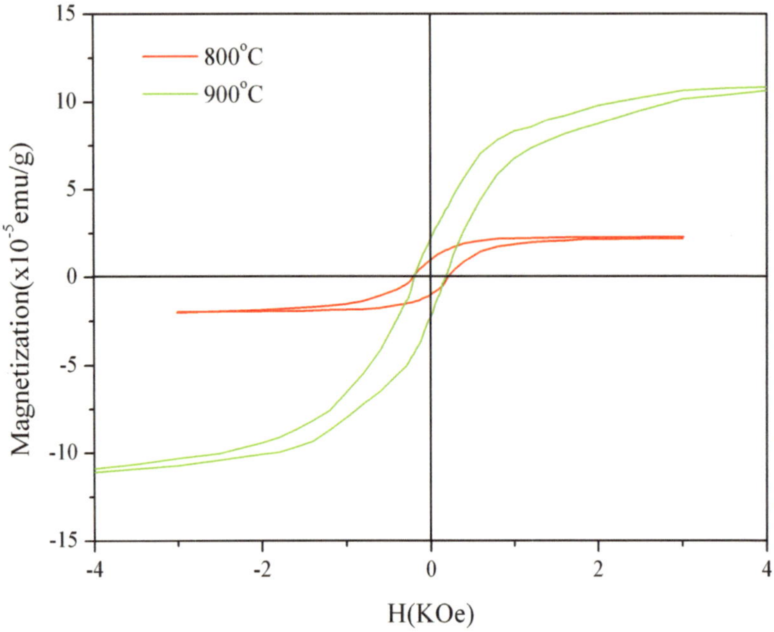

Figure 27. M-H loops at 5K of sample implanted at dose 3 × 10^{16} ions cm^{-2} and annealed [68].

Figure 28 shows magnetization as a function of temperature for Cr^+ implanted GaN at 3 ×10^{16} ions cm^{-2}, with multiple exchange interactions indicating that its decay cannot be easily fit to classical description of ferromagnetism, again in agreement with current theories concerning DMS systems with low carrier concentrations. The Cr^+ implanted GaN at 3 ×10^{16} ions cm^{-2} and annealed at 900°C showed magnetic moment at lower temperatures, retaining magnetization above the measured temperature of 380 K. This observation is consistent with the epitaxial prepared (Ga,Cr)N magnetic properties observed by Hashimoto et al. who have reported T_C higher than 400 K [35]. The value of *Ms* was higher for the sample annealed at 900°C compared to the sample annealed at 800°C. This increase is assumed to be due to the increase in Cr concentration on Ga sites. These observations suggest that annealing at 900°C is suitable for proper activation of Cr in GaN, which is also supported by the observations reported by Hwang et al. [65].

Well-saturated MH loops were observed in all samples (**Figures 26** and **27**) with finite coercivity, eliminating the likelihood of both paramagnetism and superparamagnetism [63]. Analysing hysteresis loops of the implanted samples assists in the investigation of the magnetic properties of the material. Lateral shifting of hysteresis loop was not observed, and this eliminates spin glass behaviour [64]. These observations imply that there was ferromagnetic ordering in implanted samples at room temperature. No extra peaks were observed on the XRD spectra of the implanted samples (**Figures 23**), reducing the contributions of secondary phases (Cr_xN_y, CrGa, etc.) to the observed ferromagnetism. Along with magnetization as a function of temperature measurements, FC and ZFC measurements were made on a representative sample and the data did not indicate any blocking temperature that

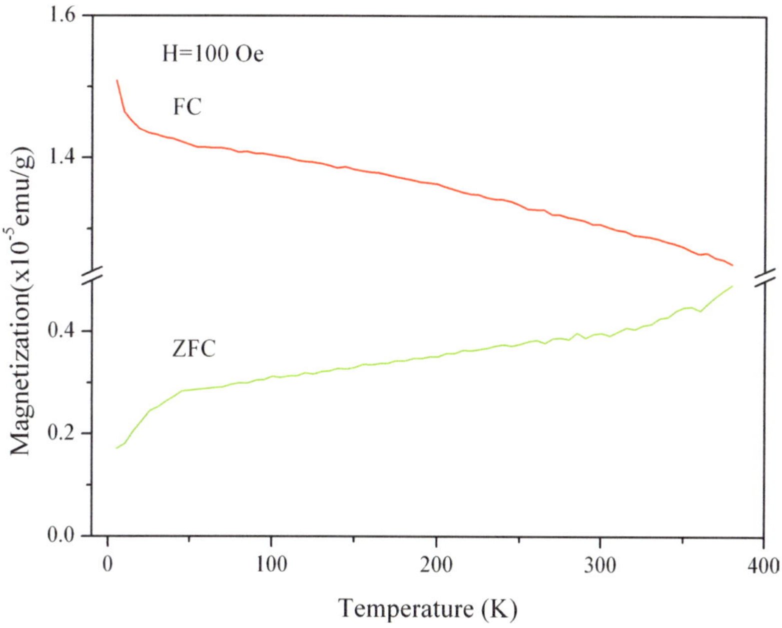

Figure 28. FC/ZFC measurements of the sample implanted at 3 × 10^{16} ions cm^{-2} and annealed at 900°C [68].

can be associated with superparamagnetic behaviour arising from undetected magnetic secondary phase clusters. **Figure 29** presents FC-ZFC measurements for Co^+ and Cr^+ ions implanted with the same dose and annealed at 900°C, which also suggests that Cr^+ ions are much suitable for the fabrication of dilute magnetic semiconductors (DMS).

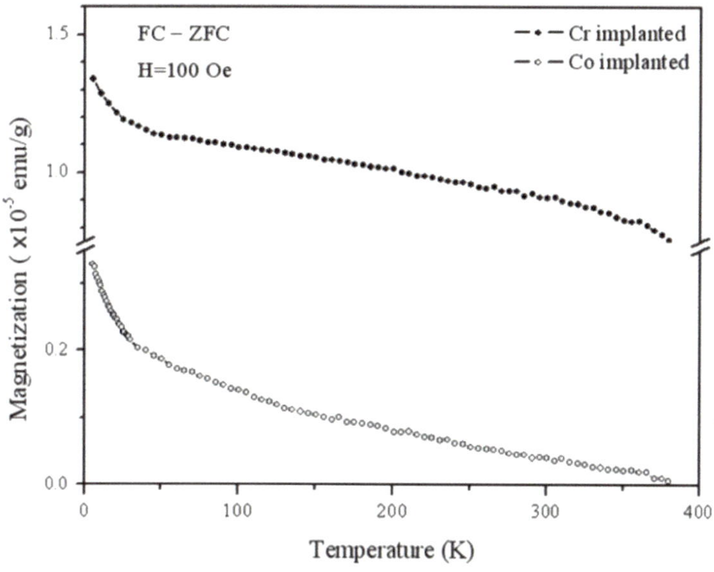

Figure 29. FC-ZFC of Co+ and Cr+ implanted samples at 3 × 10^{16} ions cm^{-2} and annealed at 900°C.

5. Conclusion

1. RBS/C measurements on the as-grown GaN samples showed good crystalline quality (χ_{min} = 1.3%) and ion-implanted samples showed recovery of lattice damage after annealing by RTA at different temperatures.

2. Typical XRD profiles were taken for Co^+ implanted GaN at doses 3×10^{16} and 5×10^{16} ions cm^{-2} and subsequently annealed at 700, 800 and 900°C and for Cr^+ implanted GaN at a dose of 3×10^{16} ions cm^{-2} and subsequently annealed at 800 and 900°C. In the as-grown samples, three main peaks appeared, corresponding to the expected diffraction from the GaN epilayer and sapphire substrate structure. Comparison of as-grown with the implanted samples at different doses showed that no secondary phase or metal-related peaks were detected in the as-implanted samples and annealed samples from typical XRD spectra. The diffraction patterns only showed the presence of peaks corresponding to the GaN layer and the substrate structure.

3. On diffraction patterns from HR-XRD scans of implanted samples, typical satellite peaks appear at lower side of the main GaN (0002) reflection. The new peaks on the HR-XRD scans are assigned to implantation-induced damage as well as the formation of $Ga_{1-x}Co_xN$ or $Ga_{1-x}Cr_xN$ in the implanted part of the samples. The lattice constants of (Ga,Co)N and (Ga,Cr)N change due to the presence of Co and Cr, respectively, suggesting that the locations of the new peaks depend on the cobalt or chromium content in the material. The shift of additional peaks towards the high angle side with annealing points to lattice recovery and improvement in the uniformity of GaCoN or GaCrN which may be due to increase in the substitution probability of cobalt or chromium atoms, respectively.

4. Well-defined hysteresis loops were observed even at room temperature by AGM in all implanted and subsequently annealed samples at different temperatures, eliminating the possibility of paramagnetism and superparamagnetism. The analysis of hysteresis loops for the implanted samples helps us to explore the magnetic properties of the material. No lateral shift of hysteresis loops was observed, which, therefore, exclude spin glass behaviour. These findings suggest the presence of ferromagnetic ordering in our implanted samples at room temperature. The room temperature coercivity H_c measured by AGM was about 100 Oe for all implanted and subsequently annealed samples.

5. The coercivity (H_c) at 5 K measured by SQUID for Co^+ implanted at 3×10^{16} ions cm^{-2} and subsequently annealed GaN samples were observed up to 275 Oe and that at 5×10^{16} ions cm^{-2} and subsequently annealed samples were observed up to 600 Oe. For Cr^+ implanted at 3×10^{16} ions cm^{-2} and subsequently annealed samples, the coercivity was observed up to 175 Oe.

6. Comparing Cr^+ implanted and Co^+ implanted samples at the same amount of dose, we observe that saturation magnetization *Ms* values are almost the same in all samples at room temperature. But at 5 K, the saturation magnetization *Ms* values are almost similar only for the samples annealed at 800°C. The *Ms* value for Cr^+ implanted and subsequently

annealed sample at 900°C is 10.7 (× 10^{-5} emug^{-1}) while that for Co^+ implanted sample is about 4.5 (× 10^{-5} emug^{-1}). A higher value of *Ms* for Cr^+ implanted GaN and annealed at 900°C suggests that samples implanted with Cr^+ ions may perform better for dilute magnetic semiconductors (DMSs) compared to Co^+ implanted. Also higher values of *Ms* for implanted samples may suggest again that 900°C is a suitable annealing temperature for the activation of dopants.

7. FC and ZFC measurements by SQUID were made on Co^+ and Cr^+ implanted representative samples subsequently annealed at 900°C. The data did not indicate any blocking temperature that can be associated with superparamagnetic behaviour arising from undetected magnetic secondary phase clusters. Magnetization as a function of temperature showed the highest reported Curie temperature $T_C \sim 370$ K for Co^+ implanted GaN and highest reported T_C above the measured temperature (380 K) for Cr^+ implanted GaN. These findings are the highest reported Curie temperatures (T_C) for Co^+ and Cr^+ implanted GaN diluted magnetic semiconductors (DMSs).

Acknowledgements

Morgan Madhuku acknowledges iThemba LABS and the National Research Foundation of South Africa for financial support.

Author details

Ghulam Husnain[1,2] and Morgan Madhuku[1,2]*

*Address all correspondence to: morgan.madhuku@gmail.com

1 National Centre for Physics, Islamabad, Pakistan

2 iThemba LABS, NRF, Johannesburg, South Africa

References

[1] M.O. Manasreh, "III-nitride semiconductors: electrical, structural and defects properties". Elsevier Science, Amsterdam, Netherlands (2000).

[2] S.J. Pearton, C.R. Abernathy, G.T. Thaler, R. Frazier, F. Ren, and A.F. Hebard, "Effects of defects and doping on wide band gap ferromagnetic semiconductors." Physica B: Physics of Condensed Matter **340** 39-47 (2003).

[3] K. Kubota, Y. Kobayashi, and K. Fujimoto, "Preparation and properties of III-V nitride thin film." Journal of Applied Physics **66** 2984 (1989).

[4] S. Kucheyev, J. Williams, and S. Pearton, "Ion implantation into GaN." Materials Science & Engineering R **33** (2-3) 51-108 (2001).

[5] S.J. Pearton, C.B. Vartuli, J.C. Zolper, C. Yuan, and R.A. Stall, "Ion implantation doping and isolation of GaN." Applied Physics Letters **67** 1435 (1995).

[6] J.M.D. Coey, M. Venkatesan, and C.B. Fitzgerald, "Donor impurity band exchange in dilute ferromagnetic oxides." Nature Materials **4** (2) 173-179 (2005).

[7] G. Schmidt and L.W. Molenkamp, "Spin injection into semiconductors, physics and experiments." Semiconductor Science and Technology **17** 310-321 (2002).

[8] I. Zutic, J. Fabian, and S. Das Sarma, "Spintronics: Fundamentals and applications." Reviews of Modern Physics **76** (2) 323-410 (2004).

[9] A.H. MacDonald, P. Schiffr, and N. Samarth, "Ferromagnetic semiconductors: moving beyond (Ga, Mn) As." Nature Materials **4**, 195-202 (2005).

[10] J.K. Furdyna, "Diluted magnetic semiconductors." Journal Applied Physics **64** R29 (1988).

[11] S.J. Pearton, C.R. Abernathy, G.T. Thaler, R.M. Frazier, D.P. Norton, F. Ren, Y.D. Park, J.M. Zavada, I.A. Buyanova, and W.M. Chen, "Wide bandgap GaN-based semiconductors for spintronics." Journal of Physics Condensed Matter **16** (7) 209-245 (2004).

[12] T. Dietl, H. Ohno, F. Matsukura, J. Cibert, and D. Ferrand, "Zener model description of ferromagnetism in zinc-blende magnetic semiconductors." Science **287** (5455) 1019 (2000).

[13] M.L. Reed, N.A. El-Masry, H.H. Stadelmaier, M.K. Ritums, M.J. Reed, C.A. Parker, J.C. Roberts, and S.M. Bedair, "Room temperature ferromagnetic properties of (Ga, Mn) N." Applied Physics Letters **79** 3473 (2001).

[14] G. Thaler, M. Overberg, B. Gila, R. Frazier, C. Abernathy, S. Pearton, J. Lee, S. Lee, Y. Park, and Z. Khim, "Magnetic properties of n-GaMnN thin films." Applied Physics Letters **80** 3964 (2002).

[15] S. Sonoda, S. Shimizu, T. Sasaki, Y. Yamamoto, and H. Hori, "Molecular beam epitaxy of wurtzite (Ga,Mn)N films on sapphire (0001) showing the ferromagnetic behaviour at room temperature." Journal of Crystal Growth **237-239** 1358-1362 (2002).

[16] K. Sato and H. Katayama-Yoshida, "Material design of GaN-based ferromagnetic diluted magnetic semiconductors." Japanese Journal of Applied Physics **Part 2, 40** (5B) L485-487 (2001).

[17] Y. Matsumoto, M. Murakami, T. Shono, T. Hasegawa, T. Fukumura, M. Kawasaki, P. Ahmet, T. Chikyow, S. Koshihara, and H. Koinuma, "Room-temperature ferromagnetism in transparent transition metal-doped titanium dioxide." Science **291** (5505) 854 (2001).

[18] K. Ueda, H. Tabata, and T. Kawai, "Magnetic and electric properties of transition-metal-doped ZnO films." Applied Physics Letters **79** 988 (2001).

[19] A. Hebard, R. Rairigh, J. Kelly, S. Pearton, C. Abernathy, S. Chu, and R. Wilson, "Mining for high Tc ferromagnetism in ion-implanted dilute magnetic semiconductors." Journal of Physics D, Applied Physics **37** (4) 511-517 (2004).

[20] J. Lee, J. Lim, Z. Khim, Y. Park, S. Pearton, and S. Chu, "Magnetic and structural properties of Co, Cr, V ion-implanted GaN." Journal of Applied Physics **93** 4512 (2003).

[21] A. Polyakov, N. Smirnov, A. Govorkov, N. Pashkova, J. Kim, F. Ren, M. Overberg, G. Thaler, C. Abernathy, and S. Pearton, "Electrical and optical properties of GaN films implanted with Mn and Co." Journal of Applied Physics **92** 3130 (2002).

[22] D. Norton, M. Overberg, S. Pearton, K. Pruessner, J. Budai, L. Boatner, M. Chisholm, J. Lee, Z. Khim, and Y. Park, "Ferromagnetism in cobalt-implanted ZnO." Applied Physics Letters **83** 5488 (2003).

[23] L. Xue-Chao, S. Er-Wei, C. Zhi-Zhan, Z. Hua-Wei, Z. Tao, and S. Li-Xin, "Difference in magnetic properties between Co-doped ZnO powder and thin film." Chinese Physics **16** 1770-1775 (2007).

[24] S. Von Molnar and D. Read, "Magneto-transport in magnetic compound semiconductors and metals." Journal of Magnetism and Magnetic Materials **242** 13-20 (2002).

[25] S. Pearton, C. Abernathy, M. Overberg, G. Thaler, D. Norton, N. Theodoropoulou, A. Hebard, Y. Park, F. Ren, and J. Kim, "Wide band gap ferromagnetic semiconductors and oxides." Journal of Applied Physics **93** 1 (2003).

[26] S. Pearton, C. Abernathy, D. Norton, A. Hebard, Y. Park, L. Boatner, and J. Budai, "Advances in wide bandgap materials for semiconductor spintronics." Materials Science and Engineering: R: Reports **40** (4) 137-168 (2003).

[27] A. Polyakov, N. Smirnov, A. Govorkov, and S. Pearton, "Electrical and optical properties of Fe-doped semi-insulating GaN templates." Applied Physics Letters **83** 3314 (2003).

[28] R. Vaudo, X. Xu, A. Salant, J. Malcarne, and G. Brandes, "Characteristics of semi-insulating, Fe-doped GaN substrates." Physica Status Solidi(a) **200** (1) 18-21 (2003).

[29] S. Heikman, S. Keller, S. DenBaars, and U. Mishra, "Growth of Fe doped semi-insulating GaN by metalorganic chemical vapor deposition." Applied Physics Letters **81** 439 (2002).

[30] R. Korotkov, J. Gregie, and B. Wessels, "Optical properties of the deep Mn acceptor in GaN: Mn." Applied Physics Letters **80** 1731 (2002).

[31] T. Graf, M. Gjukic, M. Brandt, M. Stutzmann, and O. Ambacher, "The Mn acceptor level in group III nitrides." Applied Physics Letters **81** 5159 (2002).

[32] A. Polyakov, A. Govorkov, N. Smirnov, N. Pashkova, G. Thaler, M. Overberg, R. Frazier, C. Abernathy, S. Pearton, and J. Kim, "Optical and electrical properties of GaMnN films grown by molecular-beam epitaxy." Journal of Applied Physics **92** 4989 (2002).

[33] A. Polyakov, N. Smirnov, A. Govorkov, N. Pashkova, A. Shlensky, S. Pearton, M. Overberg, C. Abernathy, J. Zavada, and R. Wilson, "Electrical and optical properties of Cr and Fe implanted n-GaN." Journal of Applied Physics **93** 5388 (2003).

[34] A. Polyakov, N. Smirnov, A. Govorkov, S. Pearton, and J. Zavada, "Electrical and optical properties of p-GaN films implanted with transition metal impurities." Journal of Physics: Condensed Matter **16** 2967 (2004).

[35] M. Hashimoto, Y. Zhou, M. Kanamura, and H. Asahi, "High temperature (>400 K) ferromagnetism in III-V-based diluted magnetic semiconductor GaCrN grown by ECR molecular-beam epitaxy." Solid State Communications **122** (1-2) 37-39 (2002).

[36] S. Pearton, C. Abernathy, G. Thaler, R. Frazier, F. Ren, A. Hebard, Y. Park, D. Norton, W. Tang, and M. Stavola, "Effects of defects and doping on wide band gap ferromagnetic semiconductors." Physica B: Physics of Condensed Matter **340** 39-47 (2003).

[37] JF Ziegler, JP Biersack, and U. Littark, "The Stopping and Range of Ions in Matter." Pergamon Press, New York (1985). http://www.srim.org/

[38] J. Lindhard, M. Scharff and H. Schiott "Range Concepts and Heavy Ion Ranges (Notes On Atomic Collisions, II) Mat." Fys. Medd. Dan. Vid. Selsk, 1-42 (1963).

[39] S. Rubin, T. Passell, and L. Bailey, "Chemical analysis of surfaces by nuclear methods." Analytical Chemistry **29** (5) 736-743 (1957).

[40] J. Tesmer, M. Nastasi, J. Barbour, C. Maggiore, and J. Mayer, "Handbook of Modern Ion Beam Materials Analysis." Materials Research Society, Pittburgh, PA (1995), pp. 40.

[41] G. Amsel, "CUTBA (Cleaning Up the Tower of Babel of Acronyms) in IBA." Nuclear Instruments and Methods in Phys Res-Section B Only-Beam Interact Mater Atoms **118** (1) 52-56 (1996).

[42] R. L. Fagaly, "Superconducting quantum interference device instruments and applications". Review of Scientific Instruments, 77, 101101 (2006).

[43] G. Kioseoglou, A.T. Hanbicki, and B. Jonker, "Comment on "Ferromagnetism in Cr-doped Ge" [Appl. Phys. Lett. 81, 3606 (2002)]." Applied Physics Letters **83** 2716-2717 (2003).

[44] B. Warren, "X-Ray Diffaction". Dover Publications, New York, USA (1990), pp. 99.

[45] C. Liu, B. Mensching, K. Volz, and B. Rauschenbach, "Lattice expansion of Ca and Ar ion implanted GaN." Applied Physics Letters **71** (16) 2313-2315 (1997).

[46] B. Pong, C. Pan, Y. Teng, G. Chi, W. Li, K. Lee, and C. Lee, "Structural defects and microstrain in GaN induced by Mg ion implantation." Journal of Applied Physics **83** 5992 (1998).

[47] F-R. Ding, W-H. He, A. Vantomme, B.P. Q. Zhao, K. Jacobs, and I. Moerman, "Lattice expansion induced by Zn channeled implantation in GaN." Materials Science in Semiconductor Processing **5** (6) 511-514 (2002).

[48] S. Qadri, B. Molnar, M. Yousuf, and C. Carosella, "X-ray determination of strain in ion implanted GaN." Nuclear Instruments and Methods in Physics Research, B **190** (1-4) 878-881 (2002).

[49] F. Zhang, N. Chen, X. Liu, Z. Liu, S. Yang, and C. Chai, "The magnetic and structure properties of room-temperature ferromagnetic semiconductor (Ga, Mn) N." Journal of Crystal Growth **262** (1-4) 287-289 (2004).

[50] A. Luukkainen, J. Keinonen, and M. Erola, "Density and size of neon bubbles in molybdenum, tantalum, and tantalum oxide." Physical Review B **32** (7) 4814-4817 (1985).

[51] V. Vishnyakov, S. Donnelly, and G. Carter, "The influence of impurities on the growth of helium-induced cavities in silicon." Journal of Applied Physics **94** 238 (2003).

[52] D. Follstaedt, S. Myers, J. Barbour, G. Petersen, J. Reno, L. Dawson, and S. Lee, "Formation of cavities in GaAs and InGaAs." Nuclear Instruments and Methods in Physics Research, B **160** (4) 476-498 (2000).

[53] E. Oliviero, M. David, M. Beaufort, J. Nomgaudyte, L. Pranevicius, A. Declemy, and J. Barbot, "Formation of bubbles by high dose He implantation in 4H–SiC." Journal of Applied Physics **91** 1179 (2002).

[54] M. Chicoin, S. Roorda, R.A. Masut, and P. Desjardins, "Nanocavities in He implanted InP." Journal of Applied Physics **94** (9) 6116-6121 (2003).

[55] G. Ouyang, X. Tan, M. Cai, and G. Yang, "Surface energy and shrinkage of a nanocavity." Applied Physics Letters **89** 183104 (2006).

[56] V. Joshkin, C. Parker, S. Bedair, L. Krasnobaev, J. Cuomo, R. Davis, and A. Suvkhanov, "Fine structure of near-band-edge photoluminescence in He-irradiated GaN grown on SiC." Applied Physics Letters **72** 2838 (1998).

[57] D. Alquier, C. Bongiorno, F. Roccaforte, and V. Raineri, "Interaction between dislocations and He-implantation-induced voids in GaN epitaxial layers." Applied Physics Letters **86** 211911 (2005).

[58] Y. Cui and L. Li, "Suppression of phase segregation during molecular-beam epitaxial growth of GaMnN using nitrogen–hydrogen plasma." Applied Physics Letters **80** 4139 (2002).

[59] Y. Shi, Y. Zhang, C. Jiang, D. Fu, and X. Fan, "The influence of implantation temperature on the magnetism and structure of Mn^+ implanted p-GaN films." Physica B: Physics of Condensed Matter **388** (1-2) 82-86 (2007).

[60] M.F. Wu, C.C. Chen, D. Zhu, S. Zhou, A. Vantomme, G. Langouche, B.S. Zhang, and H. Yang, "Depth Dependence of the tetragonal distortion of a GaN layer on Si (111) studied by Rutherford backscattering/channeling." Applied Physics Letters **80** 4130 (2002).

[61] D. Zhi-Bo, W. Wei, W. Kun, F. Tao, and S.D. Yao, "Tetragonal distortion of InN thin films by RBS/channeling." Chinese Physics Letters **26** 086111 (2009).

[62] L.R. Doolittle, "Algorithms for the Rapid simulation of Rutherford backscattering spectra." Nuclear Instruments and Methods in Physics Research Section B: Beam Interactions with Materials and Atoms **9** (3) 344-351 (1985).

[63] A. Majid, R. Sharif, G. Husnain, and A. Ali, "Annealing effects on the structural, optical and magnetic properties of Mn implanted GaN." Journal of Physics D: Applied Physics **42** 135401 (2009).

[64] L.M. Xu, Y.P. Yu, X.J. Xing, X.Y. Wu, and S.W. Li, "Enhancement of ferromagnetism upon thermal annealing in plasma assisted MBE grown mixed-phase Mn-doped insulating TiO_2 thin films." Applied Physics A: Materials Science & Processing **92** (2) 361-365 (2008).

[65] J. Hwang, Y. Ishida, M. Kobayashi, H. Hirata, K. Takubo, T. Mizokawa, A. Fujimori, J. Okamoto, K. Mamiya, and Y. Saito, "High-energy spectroscopic study of the III-V nitride-based diluted magnetic semiconductor $Ga_{1-x}Mn_xN$." Physical Review B **72** (8) 85216 (2005).

[66] G. Husnain, F. Tao, Y. Shu-De, "Structural and magnetic properties of Co^+ implanted n-GaN dilute magnetic semiconductors." Physica B **405** 2340-2343 (2010).

[67] G. Husnain, Yao Shu-De, Ishaq Ahmad, H. M. Rafique and Arshad Mahmood "Characterization of n-GaN dilute magnetic semiconductors by cobalt ions implantation at high-fluence." Journal of Magnetism and Magnetic Material, **324** 797-801 (2012).

[68] G. Husnain, Yao Shu-De, Ishaq Ahmad and H.M. Rafique "Structural and magnetic impact of Cr^+-implantation into GaN thin film." Solid State Sciences **14,** 735-738 (2012).

Chapter 3

MeV Electron Irradiation of Ion-Implanted Si-SiO_2 Structures

Sonia B. Kaschieva and Sergey N. Dmitriev

Additional information is available at the end of the chapter

http://dx.doi.org/10.5772/67761

Abstract

The effect of (10–25) MeV electron irradiation on Si-SiO_2 structures implanted with different ions (Ar, Si, O, B, and P) has been investigated by different methods, such as deep-level transient spectroscopy (DLTS), thermo-stimulated current (TSCM), Rutherford backscattering (RBS), and soft X-ray emission spectroscopy (SXES). It has been shown that in double-treated Si-SiO_2 structures, the defect generation by high-energy electrons depends significantly on the location of preliminary implanted ions relative to the Si-SiO_2 interface as well as on the type (n- or p-Si) of silicon wafer. SiO_2 surface roughness changes, induced by ion implantation and high-energy electron irradiation of Si-SiO_2 structures, are observed by the atomic force microscopy (AFM). Si nanoclusters in SiO_2 of ion-implanted Si-SiO_2 structures generated by MeV electron irradiation is also discussed.

Keywords: Si-SiO_2 structures, ion implantation, MeV electron irradiation, radiation defects, Si nanoclusters

1. Introduction

The method of ion implantation is being extensively used to create a controlled distribution of impurity concentration in a definite region of the semiconductor (in particular Si) heterostructures and integrated circuits. Advantages of the method include precise control of the implanted species depth and profile. A disadvantage of the method is the creation of a large number of defects due to the breaking of bonds between atoms in the SiO_2-Si interface and in a thin Si surface layer. Although the ion implantation of the semiconductor's structures is an essential part of microelectronics device technology, some of its shortcomings, such as the accompanying substantial radiation-induced defects, are still a matter of concern.

INTECH
open science | open minds

© 2017 The Author(s). Licensee InTech. This chapter is distributed under the terms of the Creative Commons Attribution License (http://creativecommons.org/licenses/by/3.0), which permits unrestricted use, distribution, and reproduction in any medium, provided the original work is properly cited. (cc) BY

Today, the use of radiation, in particular high-energy (MeV) irradiation, has become a valuable laboratory tool enabling researchers to study the generation, evolution, and annealing of the radiation-induced defects in semiconductor devices. Special attention is devoted to the radiation defects generated by ion implantation and by subsequent high-energy (MeV) electron irradiation. The results are a product of the collaboration between the Institute of Solid State Physics (Bulgarian Academy of Sciences) and the Joint Institute for Nuclear Research (JINR) (Dubna), Russia.

2. MeV electron irradiation of B^+ ion-implanted Si-SiO_2 structures

The effect of a 12 MeV electron irradiation on Si-SiO_2 structures implanted with 50 keV boron ions (at 1.5×10^{12} cm^{-2}) is studied by deep-level transient spectroscopy (DLTS) measurements. Metal Oxide Semiconductor (MOS) structures with oxide thickness 16.5 nm are used. B^+ ions with energy of 50 keV and a dose of 1.5×10^{12} ions/cm^2 were implanted through an SiO_2 layer. The ion energy induced the maximum damage well into the silicon matrix, about 120 nm away from the Si-SiO_2 interface. Al gate electrodes were then deposited using the photolithography technique. On the rear side (Ohm), Al contacts were deposited by thermal evaporation to form an MOS structure. The bombardment with 12 MeV electrons was completed from the gate side of the wafers. The irradiation was carried out on the MT-25 Microtron of the Flerov Laboratory of Nuclear Reactions at the JINR in Dubna, Russia. Each of the samples was irradiated for 10 or 60 s with a beam current from 5 to 6 μA. The sample holder was cooled during electron irradiation. **Figure 1** shows the DLTS spectra for MOS structures implanted with 50 keV boron ions (at 1.5 x 10^{12} ions/cm^2) before and after the irradiation with 12 MeV electrons. The DLTS signal as a function of the sample temperature during the measurement is shown for different electron doses. **Figure 1(a)** corresponds to the as-implanted sample;

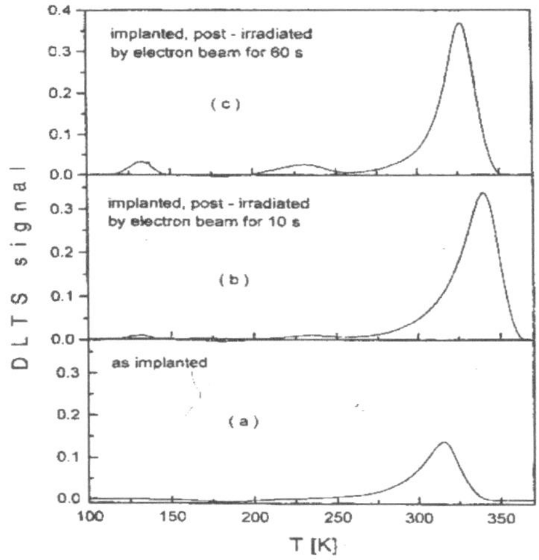

Figure 1. DLTS spectra for an Si-SiO_2 structure implanted with 50 keV B^+ before (a) and after irradiation with 12 MeV electrons for 10 (b) and 60 s (c).

Figure 1(b) and **(c)** corresponds to the implanted wafers, which were post-irradiated with electrons for 10 and 60 s, respectively.

Additional shallow levels are found in the spectra after high-energy electron irradiation and the peak intensity being dependent on the irradiation dose. The parameters as activation energy of the defects, the density of traps, and the electron-capture cross-sections are evaluated.

According to **Figure 1(a)**, one kind of defects can be associated with the ion implantation, with a corresponding peak in the spectrum at about 320 K. In **Figure 1(b)** and **(c)** curves, this maximum increased along with two more peaks, located close to 135 K and 230 K. High-energy electrons create three different defect levels at E_c–0.21, E_c–0.32, and E_c–0.40, which correspond to three distinct peaks of the DLTS spectra. The peak at E_c–0.40 coincides with that of the DLTS spectrum for the as-implanted sample (**Figure 1a**). The density of all defects increases with the increase in electron irradiation doses. The trap density of the damage related to the implantation also increases after the electron bombardment. The concentration of the two new types of defects is increased lower than the increase in the concentration of those associated with ion implantation only.

The DLTS characteristic demonstrates that 50 keV boron ion implantation (at a dose of 1.2×10^{12} ions/cm^2) creates mainly one kind of defect in the Metal Oxide Semiconductor (MOS) structures considered. These defects are located near the maximum of the ion-depth distribution, away from the Si-SiO_2 interface. High-energy irradiation by electrons enhances the contribution of such defects in the Si substrate and creates two additional states that are shallower on the energy scale. The DLTS measurements at different biases show that all states are most probably located at the Si-SiO_2 interface. The main level at E_c–0.40 eV in the Si forbidden gap band can be attributed to the phosphorus-vacancy pair (P-V) or E center. E_c–0.21 eV is related to di-vacancies and there are not enough data to relate the E_c–0.32 eV level well enough to a specific type of defect [1].

The result suggests that the high-energy electrons generate the electrically active defects at the Si-SiO_2 interface of MOS structures more efficiently than in the silicon wafer.

3. MeV electron irradiation of O^+-, P^+-, or Si^+-implanted Si-SiO_2 structures

In this work, we used Rutherford backscattering spectroscopy (RBS) in combination with channeling backscattering spectrometry (RBS/C). O^+ or P^+ and Si^+ ion-implanted Si-SiO_2 structures irradiated with high-energy electrons have been studied.

n-type Si <100> oriented wafers oxidized at 1050°C in dry O_2 + 8% HCl ambient up to 22 nm were used. The Si-SiO_2 structures were implanted by P^+ or O^+ ions with an energy of 15 keV and a dose of 10^{12} cm^{-2}. An ion beam energy of 15 keV was chosen so that the maximum number of implanted ions would be deposited close to or at the Si-SiO_2 interface. After implantation, both groups of implanted samples (with P^+ or O^+) were irradiated with different doses of 20 MeV electrons at a flux of about 10^{13} el/cm^2.s at room temperature. Three sets of samples were prepared from each of the two groups. The first set of Si-SiO_2 samples was implanted by P^+ or O^+ ions, the second set of samples was implanted and then irradiated with 20 MeV electrons with a dose of 2.5 × 10^{13} el/cm^2, and the third set was implanted and then irradiated with a dose of 1 × 10^{15} el/cm^2.

Figure 2 shows that after both doses of electron irradiation, no significant changes in the RBS spectra of the samples are observed. The first peak of the RBS/C spectrum presents the oxygen concentration and the second one, the silicon concentration at the Si-SiO_2 interface of the sample. It is obvious that the two peaks appearing after the electron irradiation almost repeat the corresponding peaks of the as-implanted sample. Obviously, the silicon and oxygen distribution at the Si-SiO_2 interface in phosphorus-implanted Si-SiO_2 samples does not change after the electron irradiation. It is assumed that the low dose (10^{12} cm^{-2}) and the type of ion implantation (phosphorus) are the reasons for these negligible changes in the Si and O's peak height.

The RBS/C spectra of oxygen-implanted and electron-irradiated Si-SiO_2 samples are summarized in **Figure 3**. It shows that the silicon and oxygen peaks in the RBS/C spectrum of

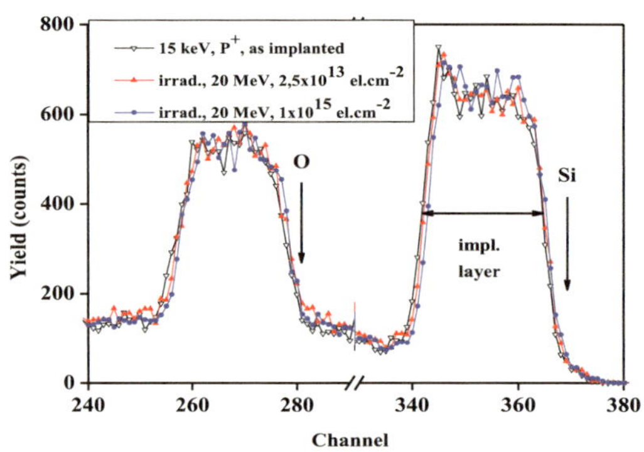

Figure 2. RBS spectra of P^+-implanted n-Si-SiO_2 samples with a dose of 10^{12} cm^{-2} (curve 1) and MeV electron irradiated with doses of 2.5 × 10^{13} el/cm^2 (curves 2) and 1 × 10^{15} el/cm^2 (curves 3).

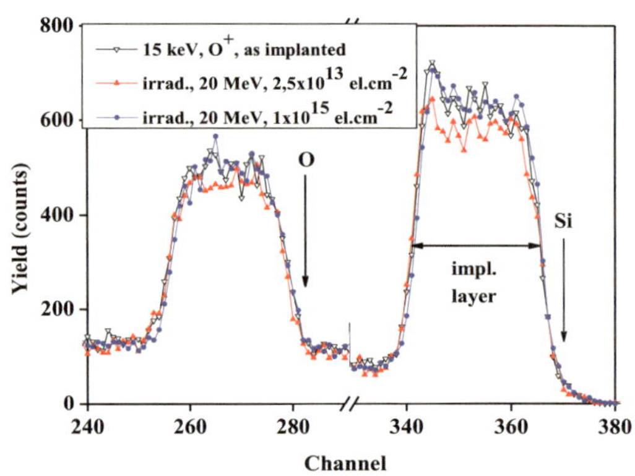

Figure 3. RBS spectra of n-Si-SiO_2 samples implanted by O^+ ions with a dose of 10^{12} cm^{-2} (curve 1) and MeV electron irradiated with doses of 2.5 × 10^{13} el/cm^2 (curve 2) and 1 × 10^{15} el/cm^2 (curve 3).

the oxygen-implanted samples increase after an MeV electron irradiation with a dose of 2.5×10^{13} el/cm^2 (curve 2). This increase is also observed after the second electron irradiation with a dose of 1×10^{15} el/cm^2 (curve 3). The increase of oxygen and silicon peaks' height (after high-energy electron irradiation) reveals an increase of silicon and oxygen concentrations in samples implanted with a low dose (10^{12} cm^{-2}) of oxygen ions.

The results demonstrate that the same doses of electron irradiation result in an increase of both oxygen and silicon concentrations at the Si-SiO_2 interface of oxygen-implanted samples. Comparing the silicon peak's increase with that of the oxygen peak in the RBS/C spectra, one can conclude that the amount of silicon accumulated is larger than that of oxygen. It has been assumed that, along with the generation of radiation defects during electron irradiation, oxygen diffusion to the silicon substrate also takes place.

RBS spectra demonstrate that the MeV electron irradiation increases the oxygen and silicon concentrations in Si-SiO_2 samples implanted by O^+ ions only [2].

4. MeV electron irradiation of Si^+-implanted (with different doses) Si-SiO_2 structures

n-type Si <100> oriented wafers with a resistivity of 4.2 Ω cm were oxidized at 1000°C up to 36 nm. The samples were implanted with Si^+ ions with an energy of 15 keV and a dose of 10^{12} cm^{-2} or 10^{16} cm^{-2}. The energy of 15 keV was chosen so that the maximum number of the implanted ions would be deposited at the Si-SiO_2 interface. Both groups were irradiated with 20 MeV electrons for different durations (60 or 120 s). After the electron irradiation for 60 s, the RBS spectra of the sample remain unchanged. After 120 s of electron irradiation, the Si peak becomes wider to the left, and the peak corresponding to O atoms increases in the same way (about 11–12 A). We assumed that the ion implantation defects, whose maximum is located close to the Si-SiO_2 interface, generate a thin layer of amorphous silicon at the silicon surface, close to the Si-SiO_2 interface. The subsequent electron irradiation generates additional radiation defects, which is the reason for the radiation-enhanced diffusion of oxygen and an insignificant increase of the amorphous layer's depth. Since after the electron irradiation both peaks of the RBS/C spectra (oxygen and silicon) widen in the same way, we can assume that during electron irradiation, the diffused oxygen penetrates the entire depth of the new amorphous silicon layer created by the additional electron irradiation.

RBS/C spectra of Si-SiO_2 samples implanted with Si^+ ions with an energy of 15 keV and a dose of 10^{12} cm^{-2} (curve B), which are electron irradiated for 60 (curves •) and 120 s (curves ■) are presented in **Figure 4a**. RBS/C spectra of Si-SiO_2 samples implanted with silicon ions at an energy of 15 keV and a dose of 10^{16} cm^{-2} (curve B), which are electron irradiated for 60 (curves •) and 120 s (curves ■) and for 120 min (curve 0) are presented in **Figure 4b**.

The as-implanted sample exhibits an RBS spectrum similar to that in **Figure 5**. We assume that the amorphous silicon layer created by ion implantation appears only after the electron irradiation. The width increases as the ion implantation dose increases (10^{16} cm^{-2}). The 60 or

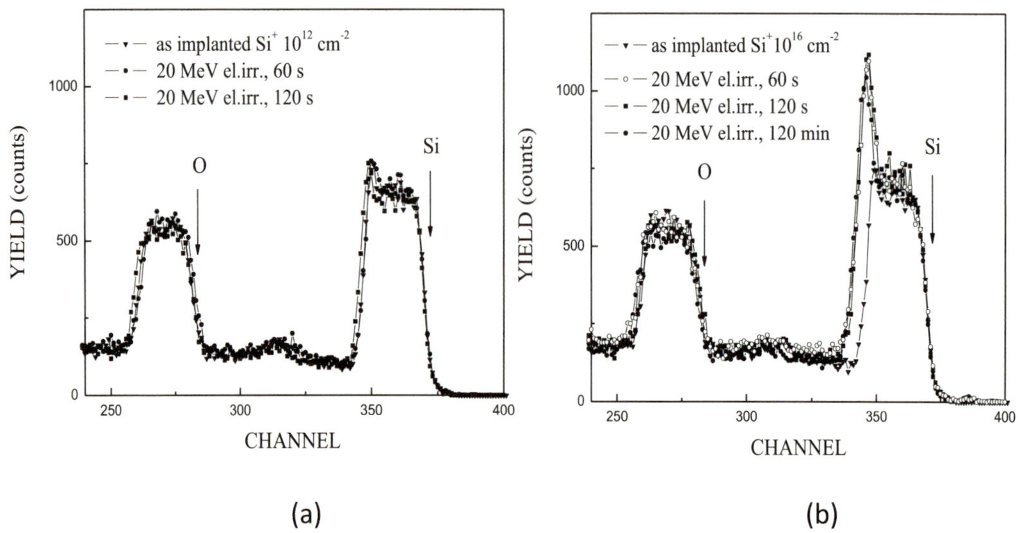

Figure 4. RBS/C spectra of Si-SiO_2 samples implanted with different doses Si^+ ions (10^{12} cm^{-2}—(a)) and (10^{16} cm^{-2}—(b)) which are 20 MeV electron irradiated.

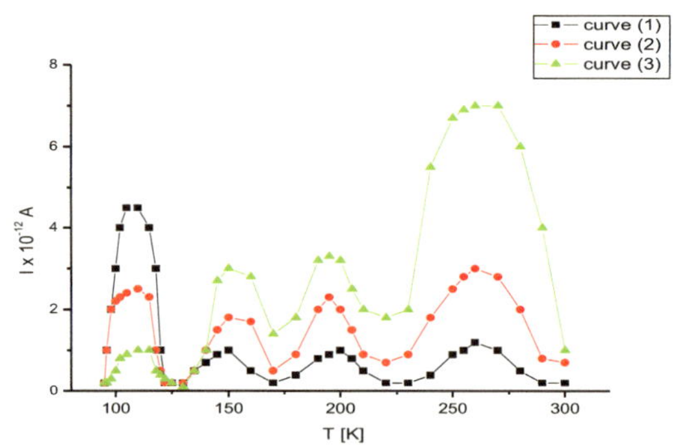

Figure 5. TSC spectra of only electron-irradiated MOS structures. Samples irradiated with doses of 1.2×10^{16} el/cm^2 (curve 1), 2.4×10^{16} el/cm^2 (curve 2), and 6.0×10^{16} el/cm^2 (curve 3).

120 s electron irradiation leads to the appearance of a new peak in the spectra. The creation of an additional amorphous silicon layer under the SiO_2 layer as a result of electron irradiation (6×10^{15} cm^{-2}) is observed. After a 120-min electron irradiation, an additional widening of the main silicon peak in the RBS spectrum is observed to the left. The height of the oxygen peak decreases insignificantly, which means that an insignificant decrease of the oxygen concentration takes place. It is demonstrated that the amount of oxygen accumulated at the Si-SiO_2 interface is proportional to the decrease of the oxygen peak in the RBS/C spectrum. Along with the generation of radiation defects, oxygen diffusion to the amorphous layer is also observed during MeV electron irradiation. One can conclude that the widening of the amorphous layer is larger (about 47 A) than the increase of the oxygen diffusion depth (which is about 12–13 A). This indicates that during electron irradiation, only a fraction of the oxygen

is able to reach the amorphous silicon layer and remain there. Obviously, the defects created by electron radiation stimulate oxygen radiation-enhanced diffusion from the surface to the sample depth. We assume that only a part of the new amorphous layer created by electrons is a result of this diffusion. It is assumed that the shift of the oxygen peak in the RBS/C spectrum by one or two channels is demonstrated. The dramatic change in the Si profile after high-energy electron irradiation for 60 s is observed at the RBS/C spectra. This profile does not change significantly after the electron irradiation with twice as high a dose (after 120 s of irradiation). A small dose of electron irradiation is sufficient to create additional radiation defects and increase the amorphous silicon layer created by ion implantation. Obviously, the radiation defects created by the additional electron irradiation are not able to significantly increase this amorphous layer. We believe that the radiation defects induced by high-energy electron irradiation interact with those already created by a previous high dose of Si ion implantation, thus creating suitable conditions for an increase of the amorphous layer. The increase of the amorphous silicon layer is a result of the structural changes at the silicon surface created by the subsequent 20 MeV electron irradiation. We assume that a low dose of electron radiation is sufficient to create favorable conditions for an additional amorphization of the silicon surfaces. An additional increase of electron irradiation (up to 120 min) induces additional breaking of the strained bonds at the silicon surface. But the substantial energy deposited cannot significantly promote the amorphization process of the silicon wafer. In this case, the results show that the amorphization process depends more effectively on the Si implantation dose than on the electron irradiation dose [3].

5. MeV electron irradiation of Ar^+-implanted Si-SiO_2 structures studied by TSC

The influence of 23 MeV electron irradiation (with doses of 1.2×10^{16} el/cm^2, 2.4×10^{16} el/cm^2, and 6.0×10^{16} el/cm^2 on the interface states of argon-implanted thin oxide Si-SiO_2 samples) has been studied by the thermally stimulated current (TSC) method. The samples are fabricated on n-Si wafers. The thermal oxidation is performed at 950°C to produce an oxide thickness of 18 nm. The samples are cooled in oxygen. Oxide thickness is measured by the ellipsometry method. Some of the Si-SiO_2 samples are implanted through the dioxide by 20 keV Ar^+ ions with a dose of 5×10^{12} cm^{-2}. Energy of 20 keV is chosen so that the ions could penetrate through the oxide with a maximum damage production close to the Si-SiO_2 interface. Al gate electrodes are photolithographically defined. MOS capacitors are performed. The samples are simultaneously irradiated with 23 MeV electrons at a flux of about 2×10^{13} el/cm^{2}s. Samples are irradiated at room temperature with doses of 1.2×10^{16} el/cm^2, 2.4×10^{16} el/cm^2, and 6.0×10^{16} el/cm^2, respectively.

The spectra reveal four peaks (at 120, 160, 195, and 260 K) corresponding to four kinds of energy states located at the Si-SiO_2 interface. Parameters of these interface states are determined by the initial rise plot and Grossweiner's methods. The interface states are associated with four discrete energy states in the Si band gap. Activation energy of traps is evaluated as: E_c–0.22 eV, E_c–0.26 eV, E_c–0.32 eV, and E_c–0.45 eV. The nature of defects corresponding to these traps is determined. There is a difference in the behavior of the first peak (which reduces its height significantly) and the rest of the ones (which increase) during electron irradiation.

A decrease in height and area enclosed by the first peak is observed which means that the density of the defect (similar to acceptor level of the di-vacancies) decreases from 5.7 to 1.1×10^{12} cm^{-2} during irradiation. The area enclosed by deeper levels in the Si forbidden gap (E_c–0.26 eV, E_c–0.32 eV, and E_c–0.45 eV) increases with increasing electron dosage. The area enclosed by the last peak increases significantly from 3.9 up to 24×10^{12} cm^{-2}. This peak of the spectra corresponding to the level E_c–0.45 eV is associated with defects similar to vacancy-impurity complexes. They are the major kind of defects created by MeV electrons in silicon MOS structures.

The results show that Ar^+ implantation generates more kinds of defects at the Si-SiO_2 interface of MOS structures as compared to the defects generated by high-energy electron irradiation only.

Typical TSC curves for Ar^+ ion-implanted structures (curve 0), which are electron irradiated (curves 1–3), are shown in **Figure 6**. The spectra (which shift slightly to lower temperatures) are located at the temperature interval 80–200 K. Curve 0 presents as-implanted samples with 20 keV Ar^+ and a dose of 5×10^{12} cm^{-2}. Ar^+ ions were used because inert gas ions have a low diffusion coefficient in the implanted layer and do not interact with Si. Ar ions form hard radiation defects only there, where (Si-SiO_2 interface in this case) they are located. Ion-implanted and electron-irradiated MOS samples with doses of 1.2×10^{16} el/cm^2, 2.4×10^{16} el/cm^2, and 6.0×10^{16} el/cm^2 are presented by curves 1, 2, and 3, respectively. Thermal and field "cleaning" of the spectra is used. The last two peaks in the spectra of Ar-implanted samples (located at about 160 and 190 K) are clearly revealed after the electron irradiation. Energy positions of the five levels are determined by the initial rise plot and Grossweiner's methods. Curves 1–3 demonstrate that additional high-energy electron irradiation creates interface states, giving priority to deeper levels in the concentration which increases three to four times after irradiation. A considerable part of defects, induced by ion implantation, increase their concentration after increasing the electron radiation dose.

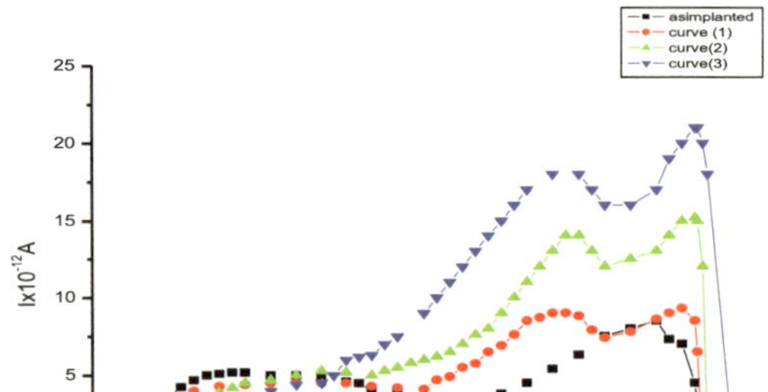

Figure 6. TSC curves for Ar^+ ion-implanted (curve 0) and electron-irradiated samples with doses of 1.2×10^{16} el/cm^2, 2.4×10^{16} el/cm^2, and 6.0×10^{16} el/cm^2 (curves 1, 2, and 3, respectively).

When Ar ion-implanted structures are irradiated with the same doses of MeV electrons, the defects due to this irradiation interact in a different way with defects previously created by Ar ions. The concentration of defects like vacancy-oxygen and di-vacancy decreases slightly, and the concentration of vacancy-impurity complexes increases as the electron irradiation dose increases. We connect the shallow-level density decrease and deeper-levels density increase with the regrouping of collected simpler defects into more complicated ones. These results showed that the defects, which contain the basic impurity of a silicon wafer, determine the radiation defect generation at the Si-SiO_2 interface during high-energy electron irradiation. As a result of interaction between MeV electrons and an Ar ion-implanted MOS structure, the concentration of radiation defects (located at the Si-SiO_2 interface) changes in a different way (depending on their energy position in the Si band gap or on their nature). The concentration of shallow traps decreases slightly after the MeV electron irradiation. The concentration of deeper traps increases with electron irradiation's dose increasing.

The defects (generated by high-energy MeV electron irradiation), which contain the basic impurity of the Si wafer of MOS structures, determine the radiation defect concentration at the Si-SiO_2 interface (the best sensitivity part of the silicon MOS structure to the radiation) [4].

6. High-energy electron irradiation of ion-implanted Si-SiO_2 structures with different oxide thicknesses

The effects of 11 MeV electron irradiation of boron ion-implanted Si-SiO_2 structures with different oxide thicknesses have been investigated by the thermally stimulated charge (TSC) method. n-type <100> Si wafer samples are oxidized to different thicknesses of 11, 25, 54, 68, 105, and 120 nm and are implanted by 20 keV boron ions with a dose of 1.2×10^{12} cm^{-2}. The ion energy and the oxide thickness of the samples are chosen so as to produce the maximum ion damage deep in the silicon substrate (for thinner oxide thickness), at the Si-SiO_2 interface (54 and 68 nm) and in the SiO_2 (for thicker oxides); Al gates are deposited onto the implanted surface of the wafers to form MOS structures. Thermally stimulated charge characteristics before and after the ion implantation or electron irradiation are measured in order to determine the parameters of the centers, induced by both treatments.

TSC spectra of 20 keV B^+ implanted samples with different oxide thicknesses irradiated with 11 MeV electrons are shown in **Figure 7**. When ions are located deep in the Si matrix, mostly one kind of defect is induced and a peak at 175 K is observed. The energy positions of the corresponding traps are determined by the slope of the initial part of the peak and by the Grossweiner's method. They are in good agreement. E_c–0.40 eV is the activation energy of these defects in the silicon forbidden gap. This center is attributed to the implantation. It is a lattice vacancy trapped at a substitutional phosphorus atom or phosphorus-vacancy pair (P-V) or "E" center. High-energy electron irradiation increases the density of these

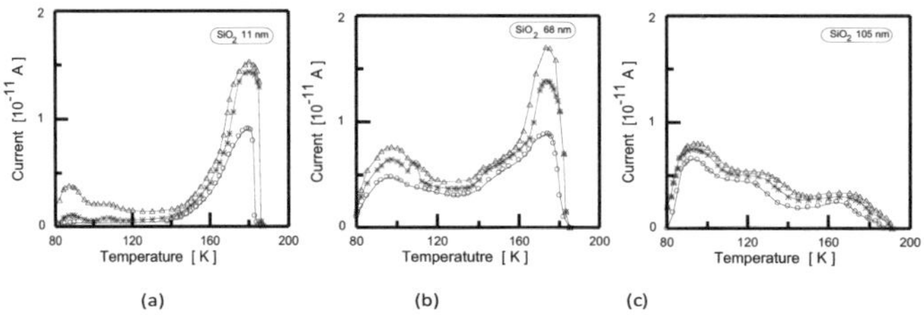

Figure 7. TSC spectra of 20 keV B^+ ion-implanted samples with different oxide thicknesses (11 nm—a, 68 nm—b, and 105 nm—c) irradiated with 11 MeV electrons. As-implanted sample with a dose of 1.2×10^{12} cm^{-2}— curve 1—(o). Implanted and high-energy electron irradiated samples for 15, 30, and 60 s are represented by curves 2 (coinciding with curves 1—o), 3 (*), and 4 (Δ), respectively.

defects created by ion implantation (E_c–0.40 eV) and additionally generates two more kinds of defects. They are shallower on the energy scale in the Si forbidden gap: E_c–0.21 eV and E_c–0.31 eV (curves 3). The density of each of them increases as the electron irradiation dose increases.

When the maximum density of boron-implanted ions is located close to the Si-SiO_2 interface of the samples, several kinds of defects are created as can be seen in **Figure 7b**. The trap energies are determined by the initial rise method. They are E_c–0.16 eV, E_c–0.21 eV, E_c–0.25 eV, E_c–0.31 eV, and E_c–0.40 eV. The nature of these defects is also determinate: shallow-level E_c–0.16 eV has been attributed to a vacancy trapped by an interstitial oxygen atom or "A" center. The next two levels E_c–0.21 eV and E_c–0.25 eV are correlated with double acceptor levels of di-vacancies. E_c–0.31 eV and E_c–0.40 eV—the last two kinds of traps—have already been discussed. In this case, the density of all levels related to the ion implantation increases with the following electron irradiation dose increasing. The density of the peaks located at 95 and 175 K (the dominant defects) increases more intensively. High-energy electron irradiation creates defects which can be associated mainly with vacancy-oxygen (A center) and vacancy-phosphorus (E center) complexes. The easier creation of interstitial atoms and vacancies near the ion-implantation complex defects are possible explanations of the intensive increase of their trap density. The process of radiation-induced defect formation as a result of both treatments is most active just in the surface region as the Si-SiO_2 interface is also a source of defect. In this case (**Figure 6b**, curves 3–4), the total density of the induced defects at the Si-SiO_2 interface (after the electron irradiation) is larger than in the case of first group samples, when the maximum damage as a result of the implanted ions is deep in the Si wafers (**Figure 6a**, curves 3–4) and in the case of third group samples (**Figure 6c**, curves 3–4), when the maximum of the implanted ions is in the SiO_2.

MeV electron irradiation of 20 keV boron ion-implanted structures generates defects, with a trap spectrum, which locates in the same temperature range as the spectrum of the as-implanted samples. After MeV electron irradiation, the density of radiation-induced interface traps depends on the disposition of the maximum of the previously implanted boron ions with respect to the Si-SiO_2 interface. The density of all levels related to the ion implantation increases uniformly with a dose increase of the following electron irradiation, in this case, when the maximum of the implantation-induced defects is located in the SiO_2 bulk.

It can be concluded that in the double treatment (B^+ implantation plus MeV electron irradiation) of the Si-SiO_2 structures, the defect generation with high-energy electron irradiation significantly depends on the location of the preliminary implanted ions relative to the Si-SiO_2 interface [5].

7. A soft X-ray emission spectroscopy study of SiO_2/Si-implanted structures irradiated with MeV electrons

The investigations were carried out with two series of samples—n-Si and p-Si wafers—cut along the <100> crystallographic plane. Oxide layers with a thickness of 150 nm (for both series) were grown in dry oxygen at 1100°C. After oxidation, the samples were cooled at a rate of about 25°C/min, in the same ambient in which the oxidation is carried out. Then, some of the samples were implanted through the dioxide with Si^+ ions, at an energy of 65 keV and a dose of 1.25×10^{12} cm^{-2}.

Prior to implantation, half of each wafer was covered with a screen and was used for control measurements to compare its properties with those of the samples exposed to ion implantation or electron irradiation or both. After implantation, both series were irradiated with 20 MeV electrons at a flux of about 10^{13} el/cm^2·s, for different durations from 15 to 120 s. The Si $L_{2,3}$ X-ray emission spectra were measured using an ultrasoft X-ray spectrometer of high spatial ($\Delta S \cdot 0.4$ μm) and energy ($\Delta E \cdot 0.4$ eV) resolution. An electron beam with an energy varying from 2 to 8 keV excited the spectra. The anode current was about 130 nA. The Si $L_{2,3}$ X-ray emission spectra of SiO_2 (150 nm)/p-Si and (150 nm)/n-Si samples implanted with Si^+ ions and then irradiated by 20 MeV electrons are shown in **Figure 8**. The spectra obtained at electron excitation energies of 2 and 4 keV are very similar to those of SiO_2. The spectra measured at 6 and 8 keV can be represented by a superposition of those of the reference specimens, SiO_2 and crystalline silicon (c-Si). The contributions of c-Si and SiO_2 to the 6 keV spectra are also shown in **Figure 8**.

Irradiation with 20 MeV electrons for 120 s leads to small increases in the SiO_2 contributions to the 6- and 8 keV spectra. The increase in the SiO_2 contribution is more pronounced for specimens prepared on n-type silicon wafers. This means that the high-energy electron irradiation leads to the oxidation of the SO_2/Si samples. This oxidation effect is more prominent for the n-type Si samples. Preliminary implantation of the samples (with 65 keV Si^+ ions) blocks the oxidation of the structures, as can be seen in **Figure 9**.

One can suggest that the oxide growth occurs due to additional number of oxygen atoms located at, or close to, the SiO_2/Si interface in the Si substrate. After irradiation, Si-Si bonds are broken. The number of oxygen atoms also increases. We show that the effect of radiation-stimulated oxidation (as a result of high-energy electron irradiation) is not strong for the SiO_2/Si samples cooled in the oxygen ambient after their thermal oxidation. We believe that a saturation of dangling bonds at the SiO_2/Si interface occurs during the post-oxidation cooling in the same ambient in which the thermal oxidation of samples was carried out. Saturation of the dangling bonds may be the reason for the increase in radiation hardness of the samples

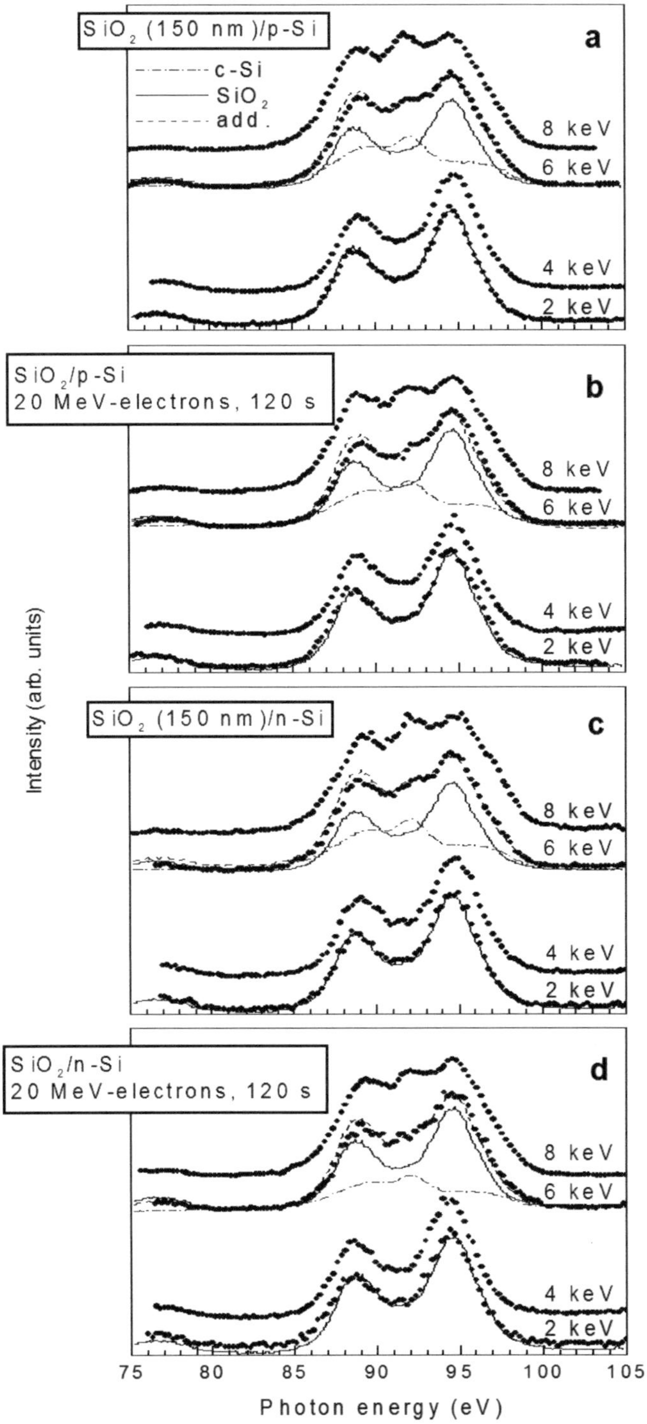

Figure 8. Si $L_{2,3}$ X-ray emission spectra measured at 2, 4, 6, and 8 keV for samples of the structure SiO_2 (150 nm)/p-Si and/n-Si as implanted (a) and (c) and after irradiation with 20 MeV electrons for 120 s (b) and (d), respectively.

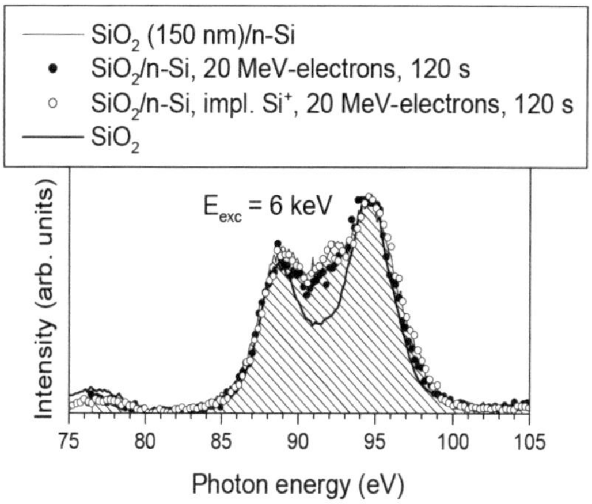

Figure 9. Si $L_{2,3}$ X-ray emission spectra for SiO_2 (150 nm)/n-Si specimens as implanted (shaded region), after 20 MeV electron irradiation for 120 s (dots), and after 65 keV Si^+ implantation and irradiation with electrons for 120 s (open symbols). The spectra were measured at an excitation energy of 6 keV. The spectrum of SiO_2 is presented as the solid line.

investigated in the present work. It was found that the effect of radiation-stimulated oxidation depends on the type of Si substrate. Our results demonstrate that the previous ion implantation blocks the oxidation process of the Si substrates, which takes place during MeV electron irradiation. This effect depends on the type of the silicon substrate, and it is more pronounced for samples prepared on n-type silicon substrates and it also depends on the post-oxidation treatment of the samples [6].

8. Si nanoclusters' generation in SiO_2 of ion-implanted Si-SiO_2 structures by MeV electron irradiation

SiO_2 surface roughness changes induced by high-energy electron irradiation of ion-implanted Si-SiO_2 structures are observed by atomic-force microscopy (AFM) measurements.

Here, amorphous SiO_2 transformation into crystalline Si is demonstrated. It is shown that silicon accumulation is observed and silicon nanocrystals are formed at the SiO_2 surface after high-energy electron irradiation. Silicon nanocrystals generated by MeV electron irradiation are observed in the SiO_2 matrix of Si-SiO_2 structures, implanted with different doses of silicon ions.

n-type CZ Si <100> wafers are cleaned and then oxidized in dry O_2 at 950°C to an oxide thickness of 20 nm. The oxide thickness is determined by the ellipsometry technique. After oxidation, the Si-SiO_2 samples are implanted through the oxide by 15 keV silicon ions with a dose of 1.5×10^{12} and 1.0×10^{16} cm^{-2}. The ion energy is chosen so as to produce maximum ion damage at the Si-SiO_2 interface. The implanted samples are simultaneously exposed to 20 MeV electrons perpendicular to the SiO_2 surface. Electron irradiation with a flux of 1.2×10^{15} el/cm^2 is carried out in a Microtron MT-25, in the Flerov Laboratory of Nuclear Reactions of JINR, Dubna, Russia.

Microscopic morphologies of SiO_2 surfaces of the implanted and irradiated Si-SiO_2 structures are observed by atomic force microscope (AFM). The AFM method is used as a measurement technique to study the changes observed in the dynamic properties of a vibrating tip that interacts with surface samples. This method allows one to resolve atomic-scale surface defects in ultra-high vacuum also. AFM images are taken over areas of 1 × 1 μm on the surface of the SiO_2-Si structures before and after the MeV electron irradiation of each of the samples.

Figure 10 presents AFM measurements of the implanted samples with doses of 10^{12} cm^{-2} Si ions (a), which is irradiated with 20 MeV electrons (b) and (c) for samples implanted with a

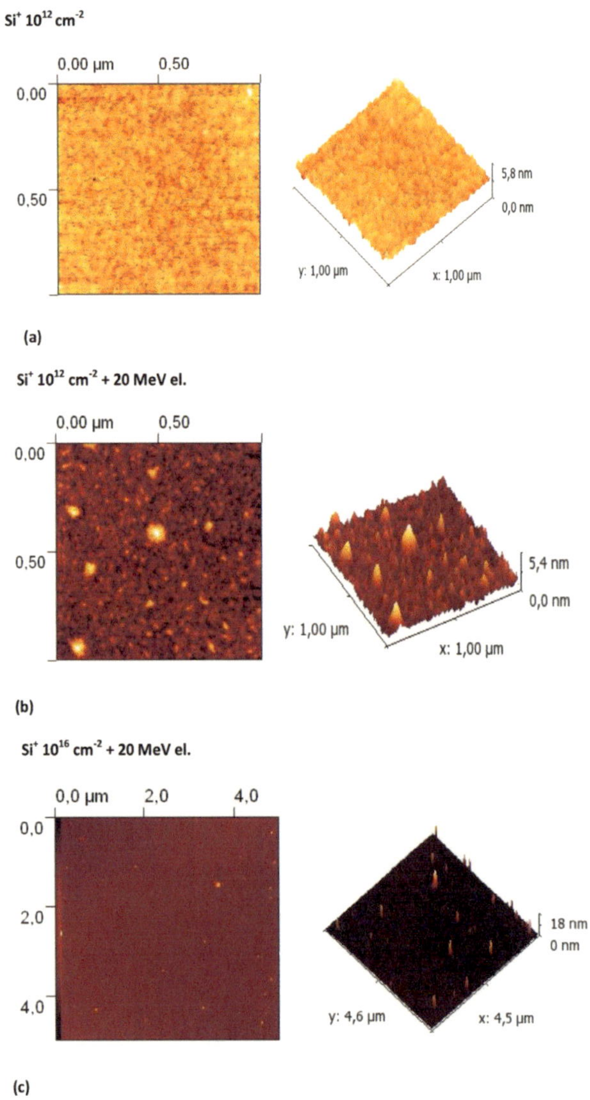

Figure 10. Surface morphology of silicon-implanted ions, Si-SiO_2 sample, before (a) and after electron irradiation (b) and (c).

dose of $10^{16}cm^{-2}$ Si ions and then irradiated with the same dose of MeV electrons. The AFM observation shows that the amorphous Si precipitations exist in the SiO_2 matrix after the ion implantation. Implanted Si atoms migrated and formed small precursor precipitations because implanted ions and implantation-induced damages were in some activated state. But it is obvious that these doses of 1.5×10^{12} and 1.0×10^{16} cm^{-2} for Si-implanted ions are not enough to create nanostructures in the oxide. Following 20 MeV electron irradiation (with a flux of 1.2×10^{15} el/cm^2) generates new silicon nanostructures in SiO_2 with a different appearance and mode depending on the previous implanted dose.

We suppose that the radiation defects (created by MeV electron irradiation) appear as a source for Si nanostructure generation in the SiO_2. We assume that these nanostructures are silicon nanocrystals generated in SiO_2 by MeV electrons. AFM observation showed that the nanocrystal size is different depending on the dose of the previous implanted ions in SiO_2. The variation of the nanocrystal size can be related to the reduction of the oxygen ion concentration at the oxide surface. The nanocrystals are bigger for samples implanted with bigger doses (1×10^{16} cm^{-2}) but their number at cm^{-2} is lower as can be seen in **Figure 1(c)** and in **Table 1**. The parameters of silicon nanocrystals created by MeV electrons in the Si-SiO_2 samples previously implanted with different doses of silicon ions are presented in **Table 1**. Obviously, the mean silicon crystal high and the mean silicon crystal length are higher for the crystals created in the samples previously implanted with higher doses of (10^{16} cm^{-2}) silicon ions.

The ion implantation breaks many of the Si-O bonds and also may knock atoms from their sites. However, most of the atoms, silicon or oxygen, bond again shortly after being dislodged. In this way, one can explain the bigger size of the nanocrystals observed after electron irradiation in silicon-implanted samples (dose 10^{16} cm^{-2}) demonstrated in **Figure 9(c)**.

MeV electron irradiation effects of surface morphology of Si-SiO_2 structures implanted with different doses of silicon ions (doses of 1.5×10^{12} or 1.0×10^{16} cm^{-2}) have been characterized by AFM observations. It has been demonstrated that MeV electrons (a flux of 1×10^{15} cm^{-2}) created Si nanocrystals in SiO_2 films in both groups of silicon-implanted samples. A form and a density of the formed Si nanocrystals depend significantly on the previously implanted silicon ion dose. The SiO_2 surface morphology has shown that Si nanocrystals are created in SiO_2 films after electron irradiation only [7].

Silicon nanocrystals generated by MeV electron irradiation in silicon ion-implanted Si-SiO_2 structures are compared with those in the SiO_2 matrix of Si-SiO_2 structures, implanted with oxygen ions (with a dose of 1.5×10^{12} cm^{-2}).

Sample	l_{mean} [nm]	h_{mean} [nm]	l_{max} [nm]	h_{max} [nm]
Si^+10^{12} + 20 MeV	50	2.3–4.2	50–124	4.2
Si^+10^{16} + 20 MeV	35	9–16	190	16

Table 1. Si crystal high and Si crystal length for the nanocrystals created by MeV electrons.

n-type Si <100> wafers are oxidized up to the oxide thickness of 20 nm. The Si-SiO_2 samples were implanted through the oxide by 15 keV Si^+ or 10 keV O^+ ions with a same dose of 1.5×10^{12} cm^{-2}. The ion energy was chosen so as to produce maximum ion damage at the Si-SiO_2 interface. The implanted samples and non-implanted samples were simultaneously exposed to 20 MeV electrons perpendicular to the SiO_2 surface. Electron irradiation with a flux of 1.2×10^{15} el/cm^2 is carried out in a Microtron MT-25 in the Flerov Laboratory of Nuclear Reactions of Joint Institute for Nuclear Research (FLNR, JINR) Dubna, Russia. The microscopic morphologies of reference sample SiO_2 surfaces, implanted or irradiated Si-SiO_2 structures were observed by the atomic force microscope (AFM) technique.

AFM images of the non-implanted sample before (a) and after 20 MeV electron irradiation (b) are demonstrated in **Figure 11**. Accumulation of nanostructures at the SiO_2 surfaces is observed only after MeV electron irradiation. Radiation defect creation by MeV electron acts as a source for Si nanostructures' generation in the SiO_2.

AFM images of Si- and O-implanted samples (a), and such samples irradiated by 20 MeV electrons (b), are presented in **Figures 10** and **11**, respectively. Energy (of 10–15 keV) and a dose (of 1.5×10^{12} cm^{-2}) for the implanted ions are not sufficient to create nanostructures in the oxide. Structural modifications on the SiO_2 surface are observed after the ion implantation. 20 MeV electron irradiation (with a flux of 1.2×10^{15} el/cm^2) generates a new silicon nanostructure in SiO_2 with a different appearance and mode. These nanostructures are silicon nanocrystals generated by MeV electrons in SiO_2. The nanocrystal sizes are different depending on the ions implanted in SiO_2.

Table 1 presents the lateral size and height of the silicon nanocrystals generated by MeV electrons in ion-implanted structures. The nanocrystal size depends on the implantation dose.

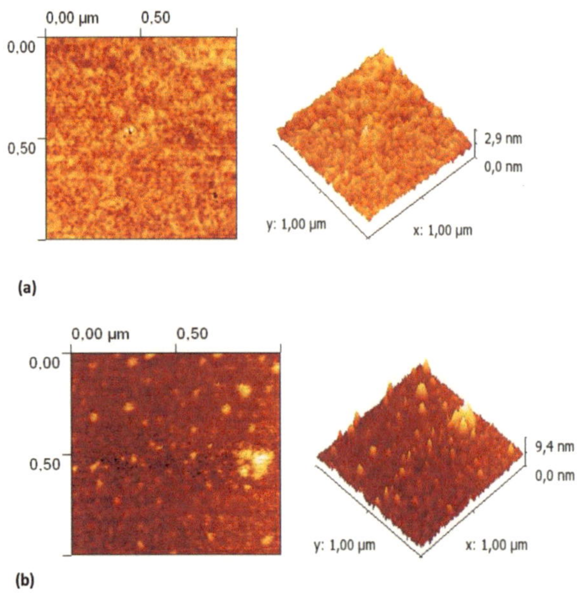

Figure 11. Surface morphology of Si-SiO_2 sample before (a) and after electron irradiation (b).

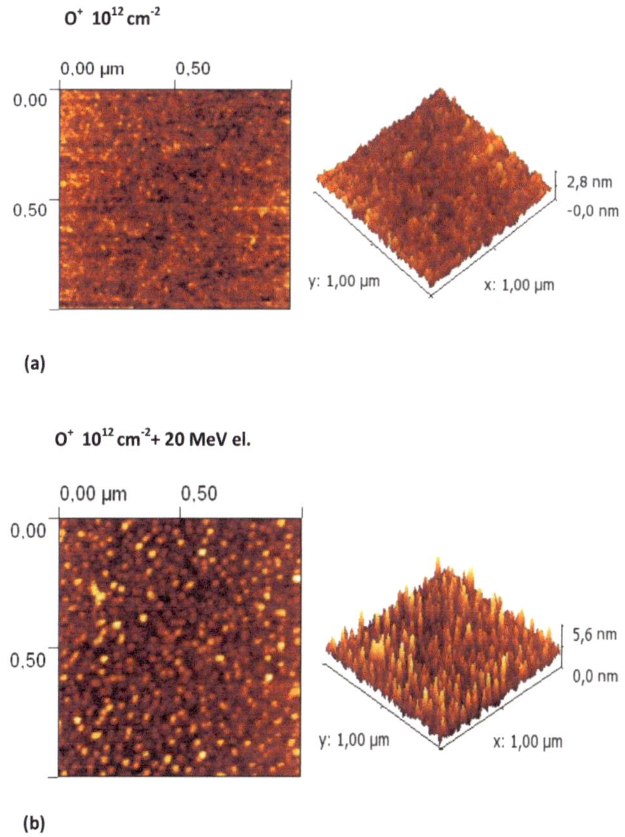

Figure 12. Si-SiO_2 sample implanted with O^+ ions (dose of 10^{12} cm^{-2}) before (a) and after (b) MeV electron irradiation.

The nanocrystal size variation can be related to the reduction of the oxygen ion concentration at the SiO_2 oxide surface. The larger nanocrystals with lower densities are in silicon-implanted samples. The nanocrystals in oxygen-implanted samples (**Figure 12**) are narrower but with a higher density. The ion implantation rearranges the Si-O-Si groups in a different form. Many of the Si-O bonds break during the implantation. It also knocks down some of atoms from their sites. But, most of the atoms (after being dislodged) bond again shortly.

An insignificant decrease of the oxygen concentrations and slight increase of silicon concentrations are observed after high-energy MeV electron irradiation. The amorphous layer's thickness on the surface of the implanted Si-SiO_2 samples also increases after MeV electron irradiation. Accumulation of silicon at the SiO_2 surface after high-energy electron irradiation is one of the reasons for silicon nanocrystals' form. MeV electron irradiation breaks Si-O bonds and free oxygen is generated. As a result, radiation-stimulated motion through defects takes place.

The surface of the implanted Si-SiO_2 before and after high-energy electron irradiation is presented in **Figure 12**. The effects of MeV electron irradiation on the surface morphology of Si-SiO_2 structures (implanted with silicon or oxygen ions) are studied by AFM. After implantation, MeV electron irradiation generates additional radiation defects in the entire Si-SiO_2 structure. It is supposed that radiation-enhanced diffusion of oxygen takes place.

This can be the reason for the creation of Si nanocrystals in the SiO_2 oxide of the implanted samples. The shape and density of Si nanocrystals depend on the type of implanted ions. Si nanocrystals are created in SiO_2 of the implanted Si-SiO_2 structure after MeV electron irradiation only [8, 9].

Our investigation into the effect of the secondary MeV electron irradiation of pre-implanted silicon heterostructures shows a variety of processes depending on post-irradiation, its energy, and doses, as well as on the ion implantation parameters. In particular, MeV electron irradiation (which when carried out at room temperature) simulates two high-temperature processes (radiation-stimulated oxidation and radiation defects annealing) in silicon heterostructures.

Author details

Sonia B. Kaschieva[1*] and Sergey N. Dmitriev[2]

*Address all correspondence to: kaschiev@issp.bas.bg

1 Institute of Solid State Physics, Bulgarian Academy of Sciences, Sofia, Bulgaria

2 Joint Institute for Nuclear Research, Dubna, Russia

References

[1] S. Kaschieva, K. Stefanov, D. Karpusov. Electron irradiation of ion-implanted n-type Si-SiO_2 structures studied by DLTS. *Appl. Phys. A.* 66: 561 (1998).

[2] S. Kaschieva, C. Angelov, S.N. Dmitriev. MeV electron irradiation of O^+ or P^+ implanted Si-SiO_2 structures. *J. Optoelectron. Adv. Mater.* 11(10): 150 (2009).

[3] S. Kaschieva, C. Angelov, S.N. Dmitriev, G. Tsutsumanova. RBS investigation of ion implanted Si-SiO_2 structures irradiated with 20 MeV electrons, *Plasma Process. Polym.* 3: 233–236 (2006).

[4] S. Kaschieva, S.N. Dmitriev. Radiation defects introduced by MeV electrons in argon implanted MOS structures. *Appl. Phys. A.* 94(2): 257–261 (2009).

[5] S. Kaschieva, S. Alexanrdova. High-energy electron irradiation of ion implanted MOS structures with different oxide thickness. *Nucl. Instr. Meth. Phys. Res. B.* 174: 324 (2001).

[6] S.N. Shamin, V.R. Galakhov, S. Kaschieva, S.N. Dmitriev, A.G. Belov. Soft X-ray emission spectroscopy of the SiO_2/Si structures irradiated with high-energy electrons. *J. Mater. Sci. Mater. Electron.* 14: 809 (2003).

[7] S. Kaschieva, A. Gushterov, P. Gushterova, S.N. Dmitriev. Si nanoclusters generated in Si-SiO_2 structures implanted with different doses of Si ions. *J. Phys. Conf. Ser.* 223: 12031 (2010).

[8] S. Kaschieva, A. Gushterov, C. Angelov, S.N. Dmitriev. Formation of Si nanocrystals in ion implanted Si-SiO_2 structures by MeV electrons. *J. Phys. Conf. Ser.* 356: 012005 (2012).

[9] S. Kaschieva, A. Gushterov, T. Apostolova, S.N. Dmitriev. *Nanoclusters in SiO_2 of ion implanted Si-SiO_2 structures generated by MeV electron irradiation*. Proceedings of the VI International Conference, "Beam Technologies and Laser Application", (2010), p. 75.

Chapter 4

Ion-Beam-Induced Defects in CMOS Technology: Methods of Study

Yanina G. Fedorenko

Additional information is available at the end of the chapter

http://dx.doi.org/10.5772/67760

Abstract

Ion implantation is a nonequilibrium doping technique, which introduces impurity atoms into a solid regardless of thermodynamic considerations. The formation of metastable alloys above the solubility limit, minimized contribution of lateral diffusion processes in device fabrication, and possibility to reach high concentrations of doping impurities can be considered as distinct advantages of ion implantation. Due to excellent controllability, uniformity, and the dose insensitive relative accuracy ion implantation has grown to be the principal doping technology used in the manufacturing of integrated circuits. Originally developed from particle accelerator technology, ion implanters operate in the energy range from tens eV to several MeV (corresponding to a few nms to several microns in depth range). Ion implantation introduces point defects in solids. Very minute concentrations of defects and impurities in semiconductors drastically alter their electrical and optical properties. This chapter presents methods of defect spectroscopy to study the defect origin and characterize the defect density of states in thin film and semiconductor interfaces. The methods considered are positron annihilation spectroscopy, electron spin resonance, and approaches for electrical characterization of semiconductor devices.

Keywords: ion beam implantation, defects, metal-oxide-semiconductor (MOS) devices, interfaces, diffusion

1. Introduction

Applications of ion implantation require an understanding of the lattice defects, which largely control the optical and electrical properties of semiconductors. Characterization techniques such as secondary ion mass spectrometry, spreading resistance, carrier and mobility profiling,

INTECH
open science | open minds

© 2017 The Author(s). Licensee InTech. This chapter is distributed under the terms of the Creative Commons Attribution License (http://creativecommons.org/licenses/by/3.0), which permits unrestricted use, distribution, and reproduction in any medium, provided the original work is properly cited. (cc) BY

Rutherford backscattering, ion channeling, and transmission electron microscopy with examples of using these techniques to investigate the dopant distribution in the implanted samples, characterize dopants that are electrically active, examine accumulation of the ion beam induced defects, and resolve their structure have been reviewed in the literature [1]. As the main feature of ion implantation is the formation of point defects in the energetic ion collisions, it is natural to present additional methods employed in semiconductor research to study atomic origin and electrical activity of technologically relevant imperfections. The prime attention will be given to characterization techniques invented in technological development of the Si/SiO_2 system, though examples of other materials systems, which can be studied by application of positron annihilation spectroscopy, electron spin resonance spectroscopy, and (photo) electrical methods are provided.

2. Positron annihilation spectroscopy

Positron annihilation spectroscopy (PAS) is now a well-established tool to characterize electronic and defect properties of bulk solids, thin films, and surfaces. PAS allows studying the electronic structure of defects in solids. The imperfections are represented by small volume defects such as vacancies, vacancy clusters, and free volume defects. Positron beams can be applied to study defects in metals, semiconductors, composite materials, and thin film systems of different crystalline structure and chemical bonding. Methodologically, PAS mainly considers the three experimentally accessible dependences schematically indicated in **Figure 1**: (i) the time-dependent distribution of annihilating photons; (ii) the angular distribution of annihilating photons; and (iii) the Doppler broadening of the 0.511 MeV annihilation line. While the time-dependent distribution of photons bears information on the electron density in the vicinity of the annihilation event, the latter two photon characteristics provide information on the electron momentum distribution. The positron lifetime gives more integral information than the momentum measurements regarding the region from which the positron annihilates. In the case of a defect-containing sample, the average electron density at a defect site can be rather defect-specific. This suggests position lifetime measurements are suitable for investigating vacancy-clustering processes in rapidly quenched or (ion) irradiated materials. The momentum measurements can also yield detailed defect-specific information. The positron energy may vary allowing examination of the depth distribution of defects in solids and interfaces. Other direct experimental methods including transmission electron microscopy and atomic diffusion are less capable in detecting open volume defects located at interfaces and surfaces. The threshold defects concentration ensured by PAS is 10^{14} to $10^{15} cm^{-3}$.

The physics of positron annihilation spectroscopy has been explained in textbooks [2, 3] and research articles [4, 5]. A positron injected into a solid becomes thermalized within a few picoseconds by ionizing collisions, plasmon and electron-hole excitations, and phonon interactions. If lattice defects are present in the material, the positron can be trapped by these imperfections. Lattice imperfections (vacancies, vacancy clusters, or dislocations), open volumes, nanoclusters, and the surface states can serve as potential wells, which effectively trap positrons. Within hundreds of picoseconds, a positron in a solid annihilates with an electron yielding two gamma rays.

http://dx.doi.org/10.5772/67760

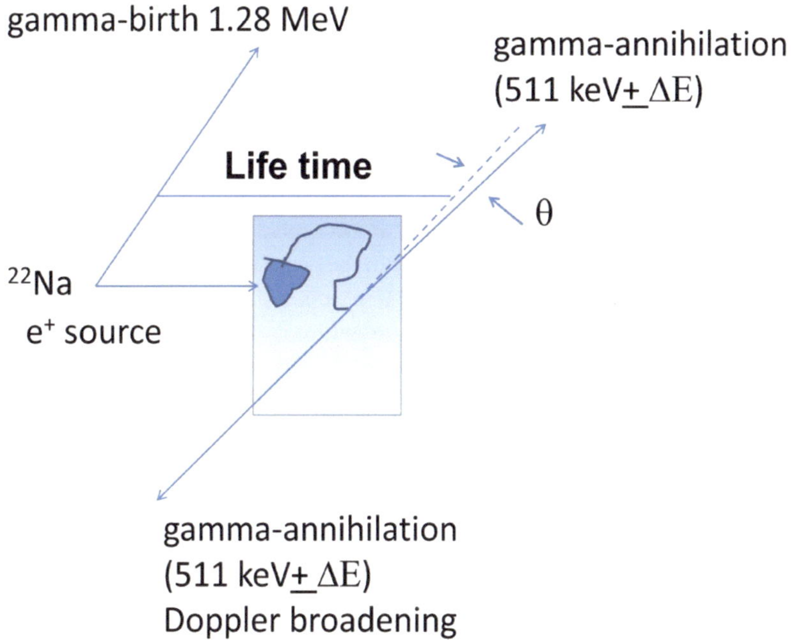

Figure 1. Schematic representation of positron annihilation indicating the basis for the three experimental techniques of positron annihilation spectroscopy: lifetime, angular correlation, and Doppler broadening.

The distribution of the Δt values for a number of these events, measured in a PAS lifetime experiment, provides the total electron density in the region of positron-electron annihilation. The positron annihilation rate λ is the reciprocal of the positron lifetime and can be described by the overlap integral of the electron $\rho^{-}(\mathbf{r})$ and positron $\rho^{+}(\mathbf{r})$ densities [4]:

$$\lambda = \pi r_0^2 c \iiint \rho^{-}(\mathbf{r})\rho^{+}(\mathbf{r})d^3 r, \tag{1}$$

where r_0 is the classical electron radius and c is the velocity of light.

Because energy and momentum are conserved in the annihilation process, the two gamma rays resulting from the electron-positron pair annihilation each have energy equal to the rest-mass energy of an electron or positron (mc^2 = 511 keV) and $\pm$ an energy increment ΔE; the two gamma rays propagate in opposite directions with some deviation θ. Since the thermal energies of the positions are about kT, the values of ΔE and θ correspond only to the momenta of the annihilating crystal electrons. The similarity of information available from Doppler-broadening spectra $P(\Delta E)$ and angular-correlation curves $N(\theta)$ can be inferred by comparing the expressions for $N(\theta)$ and $P(\Delta E)$ in terms of the independent-particle-model (IPM) probability, $R(\boldsymbol{p})$, that positron-electron annihilation yields 2γ-emission with total momentum $\boldsymbol{p}$:

$$R(\boldsymbol{p}) = \pi r_0^2 c \sum_k n_k \left| \iiint e^{-ipr}\, \Psi_{+}(r)\Psi_{-}(r) \right|^2 d^3 r, \tag{2}$$

where $\Psi_{+}(r)$ and $\Psi_{-}(r)$ are the positron and electron wave functions, respectively, n_k is the Fermi function, and k represents both the electron wave vector k and the band index. The expression for

$N(\theta)$ and $P(\Delta E)$ is represented as $N(\theta_z) = \iint R(\boldsymbol{p})dp_x dp_y$ and $P(\Delta E_x) = \iint R(\boldsymbol{p})dp_y dp_z$. The IPM approximation ignores the effects of positron-electron correlations in the solid assuming the particles act independently. The treatment of the electron-positron correlation, i.e., the enhancement of the electron density at a positron trapped by a defect site has been considered in Ref. [6]. The theory developed in this work considers the two-particle representation of an annihilating positron-electron pair. The IPM approximation is used to calculate the momentum distribution for each electron state. The individual contributions are weighted by the corresponding partial annihilation rates. The partial rates are calculated within the generalized gradient approximation. This approach was found useful when considering the momentum region where the uppermost core electron states dominate. The analysis of the momentum distribution curves up to rather large momenta becomes possible enabling identification of the chemical environment where the annihilation event has occurred. The one-dimensional momentum distribution of the annihilating electron-positron pairs can be extracted from the measurement of the Doppler broadening of the annihilation radiation. Generally, the positron-enhanced electron density can be accounted for if a constant, multiplicative factor (the enhancement factor) is used to take the many-body effects into account, although different enhancement factors must be used for valence and core electrons consistent with their degree of tight-binding.

A typical positron lifetime experiment has been described in work [7]. It can be performed by using a radioactive ^{22}Na as a positron source. The positron source material can be deposited on a sample or sealed in foil, then placed between two identical samples under study. The decaying Na nuclei emit a high energy photon at 1.2745 MeV, which is used as a start signal for the positron lifetime measurement, while a stop signal is characterized by 511 keV photons. The photons serving as start and stop signals are detected by scintillating detectors coupled with photomultiplier tubes. Detectors are chosen to optimize scintillating efficiency and resolution. The use of digitization of the detector pulses significantly simplifies the postmeasurement signal analysis. The measured positron lifetime spectrum is exponential and reveals several features such as the background noise, the time resolution, and annihilations in the source. The background noise is determined by the source activity and arises due to rapid emissions of positrons that produce false coincidences. Further, the data analysis methods are also described in Ref. [7]. Except for the least-squares fitting of the positron life time spectrum, the inverse Laplace transform and the Bayesian-probability methods have been developed. The latter two methods do not require the number of lifetime components to be *a priori* fixed and can be used if continuous lifetime distributions are expected.

The surroundings of the vacancy defect can be studied with coincidence Doppler broadening spectroscopy measurements. Nonzero electron and positron momentum causes a Doppler shift of the annihilation photons. The Doppler shift is determined by the momentum of electrons since positrons in a solid are thermalized. Analysis of the Doppler broadening of annihilation radiation provides a sensitive method of defect characterization by extracting the momentum distribution of the electrons. It allows examining high-momentum core electrons. The principle of the method lies in the analysis of the positron annihilation line shape, which directly corresponds to the distribution of momentum of electron-positron pairs as shown in **Figure 2**. The momentum itself is measured from the amount of the Doppler shift of the emitted photons. In the coincidence Doppler broadening spectroscopy developed in works

http://dx.doi.org/10.5772/67760

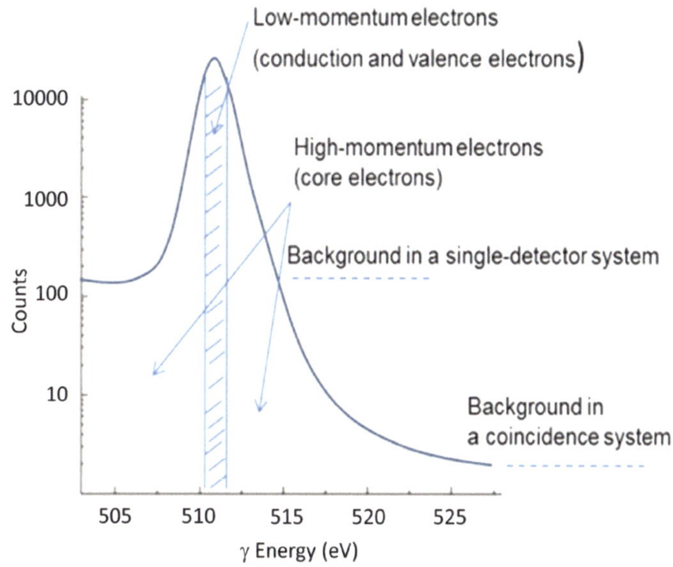

Figure 2. A typical annihilation line. After Ref. [144].

[8, 9] determination of energy of both γ rays is done simultaneously. Coincidence measurements of annihilation photons reduce the background signal by several orders of magnitude and allow observation of the high-momentum part of the spectrum, which stems from positrons annihilated by core electrons making possible identification of chemical elements surrounding a positron annihilation site.

The discovery of slow positron emitters enabled analysis of solid surfaces [10, 11]. Slow positron beams are utilized for nondestructive depth profiling of defects in surfaces and interfaces, low energy positron diffraction, and positron remission microscopy studies on surfaces. A moderator single crystal metal film (Au, Cu, W, Ta) was used to produce slow positron beams [12]. The thermalized slow positrons are emitted from the metal surface spontaneously owing to the negative positron-surface affinity. Since slow positron beam generation is a surface process, it is sensitive to surface contaminations such as carbon, oxygen, and the surface defects [13, 14]. Energy loss mechanisms and the positron processes in condensed matter are described in Ref. [15]. Except being ejected from the surface, positrons can form a positronium (P_s) by capturing a surface electron. This bound state decays from either a singlet state, p-P_s ($^1S_0$) or a triplet state, o-P_s ($^3S_1$), each having unique annihilation characteristics [16]. Positrons can become trapped by the surface states or reflected back to the interior from the surface.

When a slow positron annihilates with a core electron, the released energy can be transferred to another electron, which can be ejected and detected out of the surface. Weiss et al. [17] were first to demonstrate that a low-energy positron creates core holes through matter-antimatter annihilation generating Auger electrons with high efficiency and extremely low secondary electron background. The latter is feasible to obtain by using incident beam energy below the secondary electron emission threshold. Positron-annihilation-induced Auger-electron spectroscopy (PAES) is advantageous due to increased surface selectivity in systems where the localization of the positron at the surface causes the excitation volume to be restricted almost to the top atomic layer. In addition, because calculated PAES intensities are very sensitive to the spatial extent of the positron wave function, PAES measurements provide an important test for models of the positron surface state. This technique has been proved to be a useful tool for determining surface composition, thin film and nanocrystal characterization, and surface diffusion of atoms.

Also, positrons can be used in diffraction experiments having the advantage that interaction with solids can be easier modeled due to the sign of the scattering potential (the scattering potential between the positron and the atomic nucleus is repulsive) and the total reflection, which is only present in the positron diffraction [18, 19]. The interaction of an energetic positron with the solid may differ from that of electrons of the same energy. The differences can be associated with the relative differential and total elastic cross sections and also with the different energy loss processes for the two particles in a solid. At low energy, the inelastic mean-free path of a positron is shorter than that of an electron leading to an increased surface sensitivity of positrons. This is especially useful in examining the features of reconstructed surfaces, adsorbates, single adsorbed layers and their spacing to the substrate as well as layers with a nominal thickness in the submonolayer range. The positron scattering cross sections are marginally dependent on the specific element enabling analysis of compounds comprising unlike atoms.

High energy diffraction of positrons generates two-dimensional (2D) pattern similar to electrons, although there are several differences due to differences in the ion-core interaction and crystal potential between positrons and electrons. Kikuchi lines stemming from multiple-scattering of electrons are not observed in diffraction of high-energy positrons. The most notable feature is in the total reflection of positrons at surfaces. The positron diffraction near the critical angle is especially sensitive to the topmost atomic surface layer whereas at the critical angle for total reflection in X-ray diffraction, which is usually less than 0.2° the penetration depth of the photons into the solid still amounts to a few nanometers. Surface sensitivity of positron techniques is especially suited to near-surface measurements, which are particularly relevant to ion beam modified devices.

The technique of positron annihilation spectroscopy in conjunction with a slow positron beam has been proposed for the monitoring of ion implantation dose and uniformity [20]. Positron dosimetry can nondestructively measure doses of implanted ions with significantly higher sensitivity than that available using other techniques. The principle of the technique is that implanted thermalized positrons diffusively move in the material and become trapped by the open-volume vacancy-type defects created by ion implantation. The positron annihilation in vacancy-type defects contributes less to the Doppler broadening of the energy spectrum of annihilation γ rays compared to that in the defect-free bulk material. Doppler broadening parameter S is defined as the ratio of the number of counts in the central part of 511 keV gamma line to the total number of counts under the peak. A single parameter S describing the linewidth of the annihilation gamma ray line at 511 keV is related to the defect concentration. The concentration C of open-volume defects is related to the number ϕ of ions implanted as $C \propto \phi^{0.7}$. The defect depth profiling using positron beams has found applications in materials

http://dx.doi.org/10.5772/67760

research to study ion beam damage in both inorganic [21–23] and organic materials [24]. In the latter case, positron beam studies are of particular importance since application of X-ray or electron beams to organic materials may appear invasive [25].

Though modern MOS device technology may rely on ion-implantation free approaches [26, 27], applications of ion implantation are expanding over areas of quantum information processing [28, 29] and photovoltaics [30, 31]. Plasma immersion ion implantation enables fabrication of 3D transistor architectures [32, 33] required for scaling of metal-oxide-semiconductor field-effect transistors (MOSFETs) and is technologically more convenient for the fabrication of shallow pn-junctions. The ion implantation doping and the problems associated with the formation of point defects in the ion collision processes have been reviewed in work [34] highlighting the differences in the defect generation and accumulation in Si and Ge upon ion implantation. The dopant behavior in Ge is dominated by vacancies, while both vacancies and self-interstitials are active in Si. PAS has been applied to study point defects in interfaces between high-k dielectrics and metal [35] and Si [36]. The open volume defects were found to be located at both TiN/SiO_2 and Si/SiO_2 interfaces [37]. Annealing studies of defects indicated that while the defects in the Si/SiO_2 interface could be annealed out, the TiN/SiO_2 interface revealed an enhanced defect density due to the formation of the interfacial titanium oxynitride. Open volume defects introduced in SiN_x films [38] and SiGe/Si interfaces [39] by plasma processing have been also revealed by PAS.

3. Electron spin resonance spectroscopy

Being integral to CMOS technology, ion implantation finds its applications at the forefront of materials science for fabrication of quasi-2D materials [40, 41], exploration of electron and nuclear spins of donor atoms in silicon as qubits for quantum information processing [42], and fabrication of light-emitting diodes [43]. Pertaining to MOS device fabrication, ion implantation is known to result in generation of electron and hole-trapping centers, which are detrimental to the device performance [44]. Such trapping centers may reside in a gate oxide and its interfaces with a semiconductor and a gate electrode. In amorphous SiO_2, ion implantation induces densification and the amorphous network reconstruction, not fully consistent with the assumption of plastic deformation. Ion implantation forces SiO_2 to freeze in a nonequilibrium phase tolerating a substantial reduction in the mean Si−O−Si angle and a subsequent change in the ring distribution statistics. As such, the radiation response of SiO_2 is dependent on the intrinsic structure of the material and the incorporated strain. Possible structural modifications in amorphous SiO_2 resulting in irradiation-induced charge have been reviewed in Ref. [45]. When paramagnetic, electrically active defects can be studied by using electron spin resonance (ESR) since the method is restricted to systems with a residual electron spin. For example, molecular solids with singlet ground states are not observable by ESR. This selectivity appears as useful in research on the electronic states of conducting materials, point defects in thin films, interfaces, and nanocrystals [46–50]. For the subject of ESR describing the fundamental theory and also the primary applications of the technique one can refer to the textbooks [51, 52]. The potential of the method in application to interfaces and nanolayers is detailed in Ref. [53].

The actual quantity detected in the ESR experiment is the net magnetic moment per unit volume, the macroscopic magnetization M. The microwave absorption spectrum is described by the spin Hamiltonian consisting of two components. A spin Hamiltonian contains operators for an effective electronic spin and for nuclear spins, the external magnetic field, and parameters. Its eigenfunctions determine the allowed energy levels of the system for an ESR experiment. The characteristics of paramagnetic species are the g-value, the spin-lattice relaxation time, and the line width. The g-value is the magnitude of the electron Zeeman factor for the paramagnetic species considered. The g-value can be determined as $E = g\mu_B B$, where E is the energy of microwave, μ_B is Bohr magneton, and B is magnetic field. In the case of free electrons, the g-value becomes 2.0023. For a paramagnetic defect, the g-value is different due to the effect of local magnetic field induced by movement of electrons in their orbits. The structure of the orbits contributes to the g-values via the effect of spin-orbit coupling, which is anisotropic and depends on axis determined by the magnetic field.

The spin-lattice relaxation time characterizes interactions of a spin system with its environment and reflects the strength of the interaction between the spin system and its surroundings. The magnetic environment of an unpaired electron can give rise to the ESR line broadening. The spectral lines are broadened either homogeneously or inhomogeneously. Homogeneous line broadening can be fitted by a single Lorentzian line and indicates that all the spins are described by the same spin Hamiltonian parameters. The line width of homogeneously broadened lines depends on the relaxation time of the spins. In the case of inhomogeneous broadening, the observed signal becomes a superposition of a large ensemble of individual spin packets, which are of slightly different g-values from each other. The inhomogeneous broadening of the spectral line can be caused, for example, by anisotropy of the g-tensor or the unresolved hyperfine structure. The latter may occur when the number of hyperfine components located near nuclei is so large that the hyperfine structure cannot be clearly observed. The large line width can be also observed due to dipole-dipole interactions between the defects spins [54].

As a starting point in defect identification, it is instructive to give an overview of intrinsic and extrinsic point defects of the Si/SiO_2 system as the most comprehensively studied system in CMOS technology. Being oxidized, silicon forms network-lattice-induced dangling bond defects at the Si/SiO_2 plane. The structure of the P_b defects is dependent on the crystalline orientation of Si. The (111)Si/SiO_2 interface can be characterized by dangling bond defects of only one type—P_b centers. This is a *sp3* silicon-dangling bond directed along the [111]. The defect is of C_{3v} symmetry and can exist in four orientations in the silicon lattice [55, 56]. Thermally oxidized silicon contains the P_b density of approximately $4.9{\cdot}10^{12}$ cm^{-2}. In contrast to the (111)Si/SiO_2 interface, the (100)Si/SiO_2 interface is characterized by two ESR active defects, P_{b0} and P_{b1} as shown in **Figure 3**. When oxidation of silicon is implemented at 800–970°C, the defect density of both defect types is similar (10^{12} cm^{-2}). The P_{b1} defect is also a Si-dangling bond located slightly under the interface plane. Unlike P_{b0}, it is of monoclinic-I point symmetry [48].

The dangling bond silicon defects, the P_b centers, are often employed as sensitive probes to detect interfacial stress during the Si/SiO_2 interface formation. When Si is subjected to oxidation at $T > 900$°C, structural relaxations occur at the Si/SiO_2, and the density of P_b-centers decreases. At this point, two stages of the silicon oxidation process can be distinguished. Suboxide

http://dx.doi.org/10.5772/67760

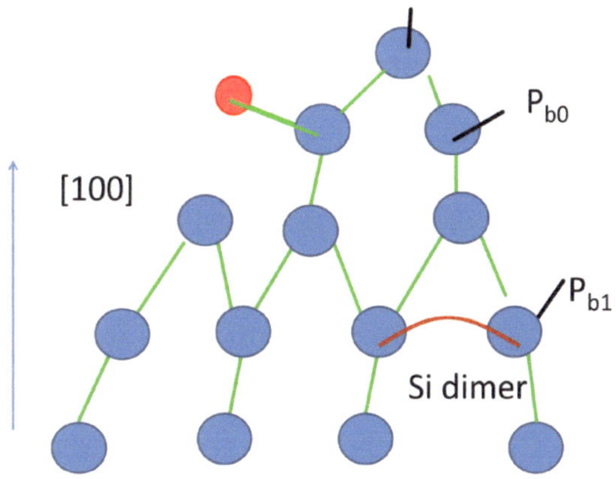

Figure 3. Schematic representation of P_{b0} и P_{b1} defects at the (100)Si/SiO_2 interface. After Ref. [145].

bonding at the Si/SiO_2 interface is diminishing when silicon is oxidizing at 850°C<T< 900°C. Increasing oxidation temperature to 1050°C reduces strain at the macroscale [57]. Spatial uniformity of the dangling bond defects is determined by the temperature conditions during silicon oxidation. ESR studies of P_b defects can be used to determine deformations at the interface from dependence of the ESR line width as a function of magnetic field angle [58, 59].

P_{b0} and P_{b1} defects in (100)Si/SiO_2 as well as P_b defects in (111)Si/SiO_2 can be passivated in molecular hydrogen [49]. Upon ion implantation or ionizing irradiation, the interface trap generation may occur. A part of the interface states appears to be due to depassivated dangling bond defects. The mechanism of the depassivation reactions has been considered within the "hydrogen model", which assumes defect precursors in SiO_2 to create mobile protons interacting with HP_b and generating P_b centers. The interface trap generation coincides with the positive charge built-up in the oxide. The model proposes that protons are introduced in SiO_2 as a product of reactions of atomic hydrogen with the hole carriers trapped in the oxide; both the atomic hydrogen and the trapped holes are produced by irradiation. It has been concluded that the positive charge trapped in the oxide is present in the form of small polarons (self-trapped holes) in amorphous SiO_2 [60]. Though in bulk vitreous SiO_2 intrinsic hole-trap centers have been found to be stable at relatively low temperatures, thin films of insulating gate dielectrics in modern MOS devices are formed by low-temperature depositions on semiconductors and could incorporate interfacial strain sufficient to support self-trapped carriers at higher temperatures. The polaronic nature of the oxide-trapped charge in amorphous SiO_2 is consistent with the recent theoretical consideration of hole and electron trapping in hafnia. The deep states of electron and hole polarons have been predicted to exist in HfO_2 with precursor sites being elongated Hf–O bonds or under-coordinated Hf and O atoms [61]. This indicates that: (i) similar mechanisms of the defect generation under irradiation or ion beam damage could be operative in MOS devices containing HfO_2 and other amorphous oxides. (ii) Dangling bond defects in oxides may not be required for the charge trapping to occur.

Of the dangling bond defects in SiO_2, there are point defects associated with a dangling bond localized either on silicon or oxygen. The *EX* center belongs to the oxygen-related defects in SiO_2. The *EX* defect is the intrinsic network-stabilized defect in SiO_2. It is formed in the upper part of the oxide when the oxidation temperature T_{ox} = 700–800°C. Being most prominent in thin oxides, *EX* is linked to the specific way thermal oxide is grown, i.e., oxidation of c-Si. As a working model, *EX* can be represented as an excess-O hole defect where an electron is delocalized over the four oxygen atoms bordering a Si vacancy [62], **Figure 4**. There are also a nonbridging oxygen hole center ($O_3{\equiv}Si{-}O{\cdot}$) [63] and a peroxide-radical ($Si{-}O{-}O{\cdot}$) [64], which are not naturally present in SiO_2 and introduced as damage defects in a postoxidation stage by irradiation with some energetic species (e.g., γ and x photons, electrons, ions).

The E' defect is also an extrinsic defect present in crystalline and amorphous SiO_2. The E' defects in SiO_2 have an unpaired electron localized at a hybrid *sp3* orbital of silicon, which is bonded to three oxygen atoms ($O_3{\equiv}Si{\cdot}$) [65]. Several schematic models of the E' centers are depicted in **Figure 5**. The model representation of E' as the bridged hole-trapping oxygen-deficiency center has not been experientially verified [66]. The model considers a paramagnetic silicon atom connected via oxygen with another silicon atom, which is the trapping center for positive charge carriers, **Figure 5(b).** Generation of E' defects may depend on hydrogen content in a-SiO_2, since dissociation energy of a strained Si–O bond by hydrogen is rather low and amounts to 0.5–1.3 eB [67]. The defect generation in interfaces and thin films by ionizing radiation or hot electron injection is sensitive to the initial content of the strain bonds in MOS devices [68]. Therefore, ESR studies could be employed to reveal the impact of the interfacial strain on the defect generation.

Since electronic devices explore charge carries in their operation, it appeared natural to establish interrelationship between the silicon-dangling bond defects and the electron states at the

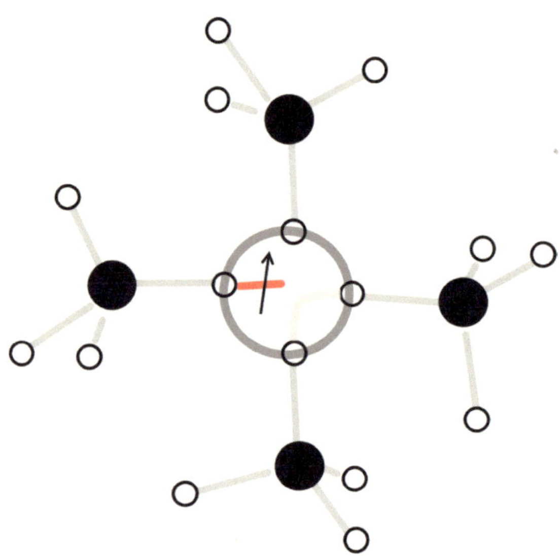

Figure 4. Schematic representation of the *EX* center.

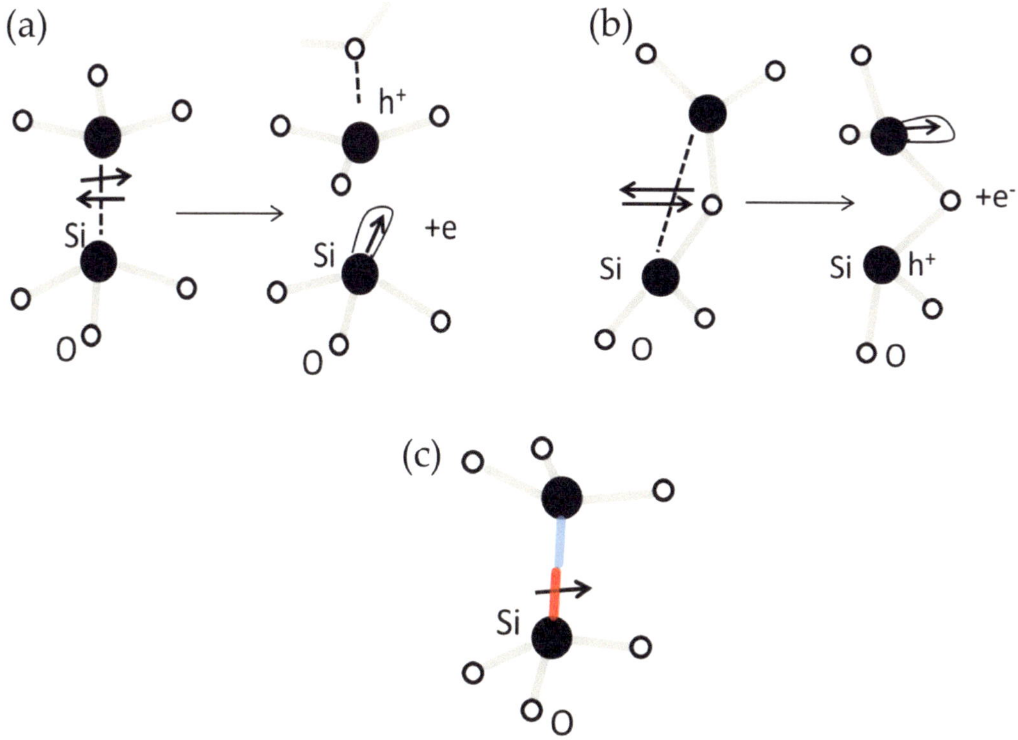

Figure 5. The first model of E′γ center (a), the model of the bridged hole-trapping oxygen-deficiency center (b), and the E′σ center model (c).

semiconductor/insulator (SI) interfaces. For the Si/SiO_2 interfaces, it is known that technology chosen for silicon oxidation is crucial for attaining low density of the interface state (D_{it}), which is directly linked to the density of silicon-dangling bonds at the Si/SiO_2 interface. The decrease in D_{it} and the P_b density was observed when steam oxidation was used to grow SiO_2. Also, the higher D_{it} values are expected at the more closely packed (111)Si surface as compared to the (100)Si one. A direct correlation between the P_b density and the free carrier concentration in the field-effect transistor channel was reported in work [69]. Further studies of electrical activity of the Si/SiO_2 defects were undertaken by using various methodologies: capacitance-voltage (CV) measurements [70], deep-level transient spectroscopy [71], and the photoionization threshold method [72]. It was firmly established that P_{b0} defects at the (100)Si/SiO_2 interface form amphoteric surface states at 0.3 and 0.8 eV above the silicon valence band edge [73]. In respect to the P_{b1} centers at the (100)Si/SiO_2, the P_{b0} и P_{b1} defect densities inferred from ESR studies were compared with the interface trap densities determined from CV measurements. It was concluded that P_{b1} does not form electrically active states within the silicon band gap [74]. Concerning the E′ center in thermal oxide, it is neutral when paramagnetic and strongly interacts with hydrogen [75]. The model for the E' center in this case is the H-terminated center denoted as $O_3{\equiv}Si{-}H$. It has been supposed that the E' center constitutes the hole trap and releases hydrogen in the form of a proton upon hole-trapping. The released proton can be trapped by the oxide network and form a donor-like surface state. When hydrogen is available in gate oxides as it can be upon an irradiation process, the neutral E' center may be again passivated serving as a hole-trapping site.

Charge trapping in gate oxides is one of the major obstacles in integration of high-k gate dielectrics in CMOS technology. Among the issues is the enhanced migration of dopant impurities originating from ion implantation steps. As such, ESR studies are indispensable to unravel point defects, which may appear detrimental for MOSFET performance. For example, ESR studies of phosphorous implanted high-k dielectrics reveal that P incorporating in the metal oxide network forms point defects by substituting for Hf or Zr in HfO_2 or ZrO_2, respectively [76]. Such defects formed due to enhanced migration of dopant impurities during dopant activation thermal steps may potentially trap charge.ESR studies have been applied to diverse ion-implanted systems. In SiO_2, a substantial reduction in *S* and E'_γ centers (Si enrichment in the oxide) was found when *in situ* ultrasound treatment was applied during implantation of Si^+ ions into thermal SiO_2 on (100)Si [77]; ESR found a radical mechanism of degradation of the ion-implanted photoresist [78]. Applications of the ESR techniques to study ion-beam-induced implantation damage in carbon-based materials have been described in Ref. [79].

ESR techniques have been explored in studies of spintronic materials fabricated by ion implantation. To probe the spin relaxation, the technique of choice is the pulse-electron spin resonance spectroscopy. ESR studies have been undertaken to measure spin relaxation times of dopants in Si. Shallow donors in Si are known for their long relaxation time suggesting a possible application of spins as qubits. The transverse relaxation time measured for isolated spins is associated with the decoherence time. ESR studies have been used to determine spin relaxation times in Sb-implanted isotopically enriched ^{28}Si [80]. It has been shown that annealing of ultralow dose antimony implants leads to high degrees of electrical dopant activation with minimal diffusion. Spin relaxation times were increased when paramagnetic defects at the Si/SiO_2 interface were passivated by hydrogen. Except for the Si/SiO_2 system, pulsed ESR experiments have been used to characterize the coherent spin dynamics of nanofabricated nitrogen vacancy centers in nitrogen implanted high-purity diamond [81].

4. Electrical characterization of semiconductor interfaces: semiconductor doping, interfacial and oxide charges

4.1. Steady-state capacitance

An overview of charge carrier profiling, steady-state and transient capacitance, deep-level transient spectroscopy methods can be found in Ref. [82]. CV methods are most frequently used to extract parameters critical for operation of semiconductor devices. The interface trap densities, the fixed oxide charge, the carrier concentration in a semiconductor, and the permittivity of an insulator can be obtained from CV measurements. Here, more emphasis is given to basic limitations of CV methods, possible errors, and examples of using CV techniques.

The measure of charge responses in MOS devices as a function of electric field is the differential capacitance. To account for the interface trap effects, the Berglund method that establishes the relation between applied voltage across the MOS structure and the band-bending in equilibrium can be used [83]. **Figure 6** exemplifies the energy distributions of the interface states at the (100)Si/SiO_2 and (100)Si/HfO_2 interfaces in panels (a) and (b), respectively. The D_{it} distributions in panel (a) are obtained using the Berglund procedure. The results of D_{it} distributions extracted from a low-frequency CV curve and *ac* admittance data are compared in panel (b) indicating a perfect match in the energy range where the fast interface states contribute to the emission of charge carriers. The Si/SiO_2 interface trap distributions derived from the 100 Hz CV curves reveal two peaks centered at 0.25 and 0.85 eV above the Si valence-band edge, **Figure 6(a)**. The peaks are superimposed on the U-shaped background corresponding to a continuous distribution of the surface states in energy and ascribed to the existence of weak Si–Si and Si–O bonds at the Si/SiO_2interface [84]. The observed peak energy positions correspond to the (+/0) and (0/−) transitions of the amphoteric P_{b0} defect. No measurable contribution of the P_{b1} center to the D_{it} could be detected in the central part of the Si bandgap, in agreement with the previous studies [74], which compare the total interface trap density N_{it} and the P_{b0} and P_{b1} densities inferred from the ESR data. The estimation of the total interface trap density N_{it} in work [74] was done according to Gray and Brown as described in work [85]. This procedure is advantageous over the low-frequency Berglund method in the following: (i) It allows detection of the interface states close (20 meV) to the Si band gap edges, inaccessible for room-temperature

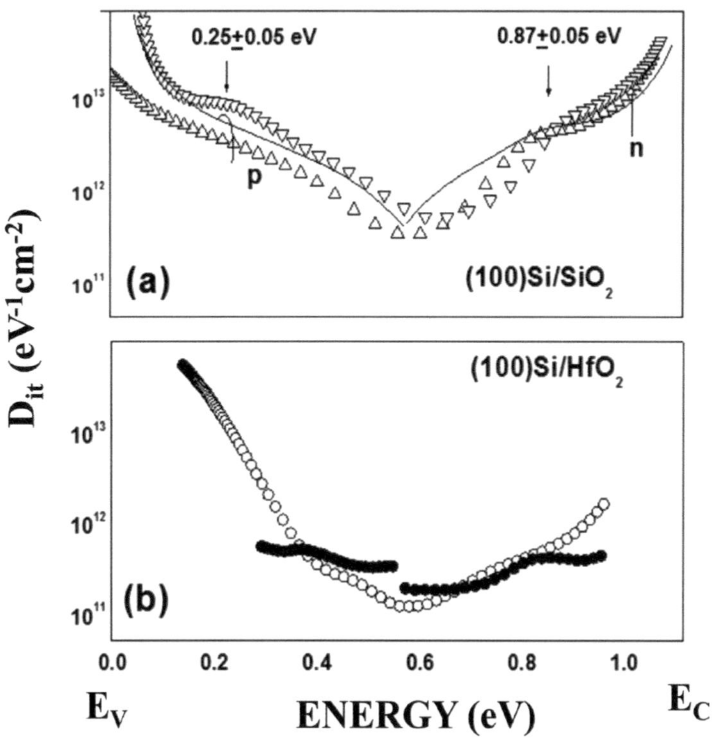

Figure 6. The interface trap distributions inferred from the capacitance-voltage measurements following the Berglund method and compared with these determined from *ac* conductance data as denoted by (•) symbols in panel (b).

CV analysis. (ii) It is decoupled from the uncertainty of Si surface potential determination near the band edges when the interface trap density is high.

A strong capacitance dispersion and dc leakage current may hamper application of CV methods based on low-high frequency CV measurements. As such, the Terman procedure based on comparison of the calculated ideal and experimental high-frequency CV curves may have limited applications in determining the interface trap densities in the case of interfaces of high-k dielectrics with semiconductors. Also, the interface trap contribution to the CV curve shift in voltage due to the interface traps becomes less prominent when equivalent electrical thickness of an insulator decreases [86, 87].

CV techniques can be used to extract the charge carrier profile in a semiconductor, the important characteristic of ion-implanted devices. The dopant profile is obtained from the high-frequency CV curve to minimize possible uncertainties stemming from the interface trap charge contribution to the depletion layer capacitance. The principle behind the dopant profiling is that as the semiconductor becomes depleted by the majority carriers, the capacitance decreases. A rapid decrease of the capacitance indicates a low dopant concentration, whereas a slow reduction indicates a high doping level. The capacitance as a function of voltage is related to the majority carrier density and can be obtained from the slope of the Mott-Schottky curve [88].

A variant of CV carrier profiling, which employs an electrochemical contact to a semiconductor, is an electrochemical capacitance-voltage (ECV) technique. ECV may appear as advantageous compared to the conventional CV methods due to its capability to measure spatial-ionized impurity distribution to practically unlimited depth, not being hindered by the breakdown at a high doping level [89]. ECV profiling can be applied to materials, which cannot be studied by the Hall measurements, for example, to conductive ferromagnetic semiconductors [90]. Despite its utility, ECV applicability is limited by the sample thickness when it is comparable with the Debye length, or, if a sample consists of several thin layers, which are either of different chemical composition or doping. The charge transfer at the interface is an important difference between a semiconductor/electrolyte (SE) interface and a Schottky contact. In the former case, it is supported by an electrochemical process. Parameters of the SE interface are determined by the electronic structure of the interface. The potential distribution in the SE interface and the effects of the semiconductor surface states on the potential redistribution between the semiconductor and the Helmholtz layer have been considered in review articles [91, 92]. When the surface states are not present at the semiconductor electrode the reverse bias drops across the semiconductor space charge region. It is than possible to determine the carrier concentration in the semiconductor. Except for the charge trapped in the surface states, there can be other charges, which result in the flat band voltage shift (V_{fb}) and modify the capacitive-frequency responses. An interfacial electric dipole layer can also result in a V_{fb} shift when the latter coincides with the change in electron affinity indicating that the dipoles attached to the semiconductor surface contribute to the V_{fb} shift, not surface charges.

Analysis of CMOS devices with nanometer thin insulators requires taking into account quantum-mechanical effects in the accumulation capacitance [93, 94] and the inversion capacitance [95] in order to extract the equivalent oxide thickness or the semiconductor doping density, respectively.

The doping density can be extracted from the inversion layer capacitance by relating the depletion layer width W_D and the carrier concentration $N_{A,D}$

$$W_D = \sqrt{\frac{4\varepsilon_s kTln(N_{A,D}/n_i)}{q \cdot (N_{A,D})}}, \tag{3}$$

where n_i is the intrinsic concentration in a semiconductor at a given temperature T and q is the elemental charge [96].

Alternatively, the doping density can be known from the band bending at the onset of strong inversion $\Psi_{s\,inv} \approx \frac{2kT}{q} \ln\left(\frac{N_{A,D}}{n_i}\right)$. The surface potential is obtained by using the Berglund integral. For the scaled MOS devices, one has to take into account the contribution of the finite density of states and the finite inversion layer thickness to the inversion layer capacitance or utilize the capacitance in the weak inversion to extract the substrate doping (cf. **Figure 2** in Ref. [95]). The inference of the semiconductor substrate doping from the inversion capacitance may appear to be superior over other experimental approaches, because it is decoupled from the possible contribution of the interface states to the depletion layer capacitance. This technique has been applied to trace boron concentration in silicon as a probe for the presence of radiolytic hydrogen in SiO_2 when analyzing the impact of vacuum ultraviolet irradiation and ion implantation of fluorine and argon on charge built-up in Si/SiO_2 MOS systems [97, 98]. Local characteristics of dopants can be obtained on semiconductor devices by using the scanning capacitance microscopy, a technique based on local capacitance-voltage analysis with submicron spatial resolution [99].

4.2. Steady-state *ac* conductance

The dynamic electrical responses of junction space-charge layers can be probed by using *ac* admittance spectroscopy or transient spectroscopy methods. These methods are applicable to both the deep bulk trap [100, 101] and interface trap [102, 103] studies in MOS devices. The *ac* admittance method is a classical approach to characterize the interface states in MOS structures [104]. The method better accentuates fast interface states, which are spatially located at the SI interface plane. The method considers the imaginary part of the measured admittance, which is directly linked to the charge trapped and emitted from the interface states as a consequence of the applied *ac* electric field. The localized states exchanging charge with the majority carrier band of a semiconductor respond to *ac* signal with both the capacitive and conductive components. At a particular frequency ω which is $\omega\tau = 1$, where τ is the characteristic time constant for the charge exchange with the localized state. The ratio G_p/ω reaches a maximum value directly proportional to the density of the surfaces states D_{it}. The trap occupancy is modulated by the semiconductor surface potential Ψ_s. The capture cross sections $\sigma_{p,n}$ and the trap densities $N_{t(p,n)}$ can be inferred from the frequency dependences of conductance exemplified in **Figure 7**. The interface trap resonances can be analyzed by using different models. Initially, it was suggested that there exists a quasicontinuous distribution of the interface states localized at the SI interface and that the surface charge and potential are uniform all over the interface. The broadening of the experimental normalized conductance curves was explained by Nicollian and Goetzberger as related to a random oxide charge and charge of the interface states distributed in the interface plane [105]. The tunnel recharging of the traps has been considered in Ref. [106]. To account for asymmetric conductance peaks, another model suggested that the interface traps at a particular energy have a range of cross sections spanning over orders of magnitude [107].

In nanoscale CMOS devices, the excessive leakage current impacts characterization of the interface traps by application of *ac* admittance spectroscopy. It has been demonstrated that errors in series resistance are critical when D_{it} values are determined at the accumulation band bending, while high tunnel currents hamper characterization of the midgap interface states [108]. The practical solution of the problem associated with the interface trap characterization in tunnel MOS-devices is the use of the charge pumping method [109, 110]. When the leakage current does not impede the interface trap analysis, the interface states in the (100)Si/SiO_2 and (100)Si/HfO_2 entities can be reliably inferred from the capacitance frequency dispersion [111, 112] or *ac* admittance spectroscopy combined with the CV methods [113]. In the latter work, it has been observed that the D_{it} density measured on Hf-containing samples subjected to a high-temperature anneal in oxygen and a subsequent passivation in hydrogen is still higher than that inferred for the equally treated (100)Si/SiO_2 interface. After passivation in molecular hydrogen, both the HfO_2 and SiO_2 interfaces with Si exhibited the D_{it} peak positioned at 0.4 eV above the silicon valence band top. When P_{b0} centers are passivated by molecular hydrogen the *ac* conductance responses are dominated by the contribution of the slow states, which are usually ascribed to the oxide-related imperfections. The slow states giving rise to the feature observed at 0.4 eV are likely to originate from the near interfacial oxide interlayer and could exist due to a lattice distortion in strained interfaces.

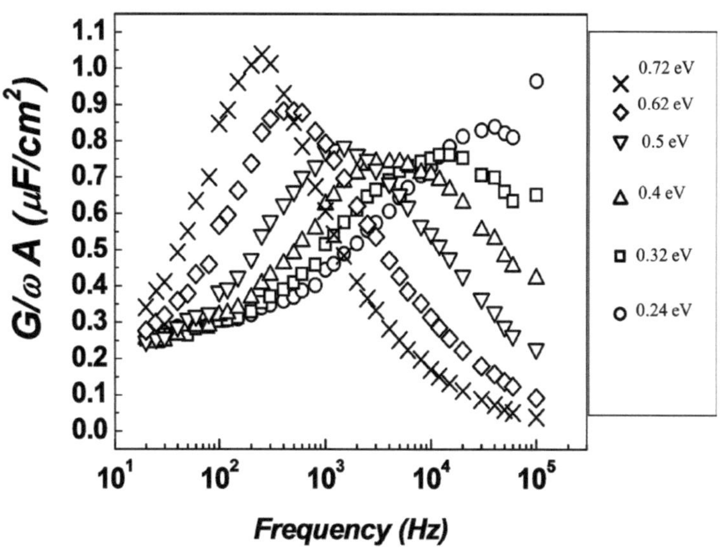

Figure 7. Equivalent parallel conductance as a function of frequency. The points are experimental values taken on a MOS-capacitor at different surface potentials ψ_s. The silicon substrate is of p-type conductivity.

4.3. Transient capacitance

Transient-capacitance spectroscopy has been initially developed to study deep bulk trap levels in semiconductors and termed by Lang as deep-level transient spectroscopy (DLTS). The capacitance DLTS is a preferred variant of the transient measurements, because it allows to separate minority and majority carrier emissions [114]. The technique is based on recording fast capacitance transients and passing the transient signal through a rate window circuit using a boxcar integrator and predefining the width of the gate pulse, the integrator response time, and the rate-window time constant. A lock-in amplifier used instead of a boxcar integrator requires settings for the rate-window, the initial gate-off period and the phase. When the traps are continuously distributed in energy (such as the interface traps) the measurement yields an emission time-constant spectrum, which depends on both the trap distribution and capture cross sections. A conventional DLTS procedure uses biases in depletion and pulsed voltage to populate interface traps with majority carriers. The responses of the device capacitance are recorded as the interface trap occupancy tends to equilibrium distribution. The energy of the traps can be determined independently of the emission rate by using two charging pulses of slightly different amplitude to selectively populate the interface traps [115]. A new method to determine capture cross sections independently of temperature and energy has been proposed in the work [116]. The method exploits the use of small trap-filling pulses to narrow the energy range within which the surface states become populated with majority carriers. Schematic diagrams representing (a) energy bands at the SI interface and (b) the pulsing sequence are shown in **Figure 8**.

When a voltage pulse sequence ΔV is superimposed on a constant voltage biasing a MOS structure to the surface depletion by the majority carriers, the capacitance difference recorded between times t_1 and t_2 is expressed as

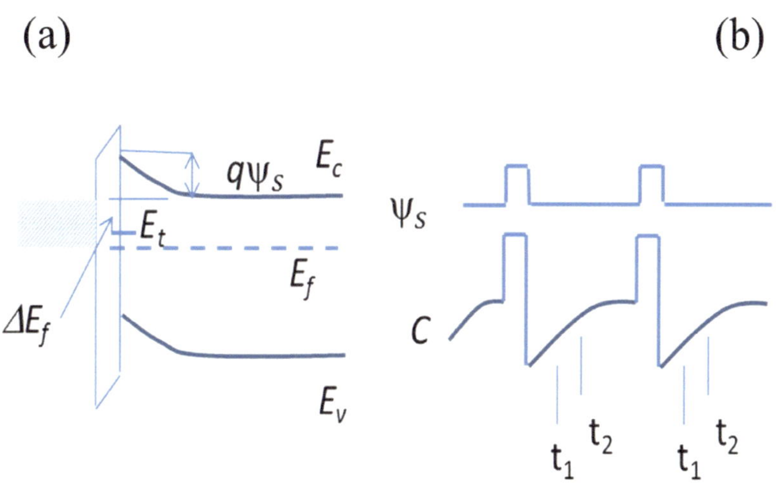

Figure 8. Schematic diagrams representing (a) energy bands at the SI interface and (b) the capacitance and the surface potential at the SI interface.

$$\Delta C = A\int_{E_v}^{E_c} N_s(E)\left[e^{-\frac{t_1}{\tau_n}} - e^{-\frac{t_2}{\tau_n}}\right][f_0(E) - f_1(E)]dE, \tag{4}$$

where $N_S(E)$ is the surface state density at energy E, τ_n is the emission time constant for electrons when considering n-type semiconductor. A constant $A = C_0^3/\varepsilon_s C_{ox} N_D$, where C_0 is a capacitance at reverse bias, ε_s is the Si permittivity, and C_{ox} is the insulator capacitance, N_D is the substrate doping. The integration limits span from the valence band edge E_v to the conductance band edge E_c, and E_f is the Fermi level.The electron occupation of the surface states at the surface potential values Ψ_s and $\Psi_s - \Delta E/q$ is described by the Fermi functions $f_o(E)$ and $f_1(E)$. As the pulse amplitude is small, the occupancy of the surface states can be approximated by the δ function, and Eq. (4) can be written as the capacitance of a discrete level.

$$\Delta C = AN_s(E_t)\left[e^{\left(-\frac{t_1}{\tau_n(E_t)}\right)} - e^{\left(-\frac{t_2}{\tau_n(E_t)}\right)}\right] \tag{5}$$

For a discrete level, DLTS spectrum peaks at

$$\tau_n = \frac{t_2 - t_1}{ln(t_2/t_1)} \tag{6}$$

The emission time constant is expressed as

$$\tau_n = [v_{th} \cdot N_c \cdot \sigma_n e^{(-\Delta E_t/kT)}]^{-1}, \tag{7}$$

where v_{th} is the thermal velocity of electrons, $N_c = N_D e^{(qV_f/kT)}$ is the effective density of states in the conduction band, σ_s is the capture cross section for electrons, and ΔE_t is the activation energy.

Assuming a capture cross section is exponentially dependent on energy

$$\sigma_n = \sigma_0 e^{-\Delta E_\sigma/kT}, \tag{8}$$

with σ_0 and ΔE_σ being the preexponential factor and the activation energy, respectively, a set of the capture cross sections at different energies can be expressed as

$$\sigma_n(E_t, T) = \sigma_0(E_t)e^{\left(-\Delta E_\sigma(E_t)/kT\right)}. \tag{9}$$

The apparent activation energy and the energy-dependent term $\sigma_0(E_t)$ can be determined from the Arrhenius plot. Repeating the DLTS measurements at different gate voltages (i.e., different surface potentials), one obtains $\sigma_0(E_t)$. The surface potential values can be determined from CV curves. The doping density and the oxide capacitance are estimated from the CV curves under the inversion and the accumulation, respectively.

In DLTS measurements, the bias dependence of the peak is a distinct signature of the charge carrier emission from the interface states [117]. Being characterized by DLTS and CV measurements, the oxide charge, the interface state densities, and capture cross sections in the energy gap can be utilized to obtain surface recombination velocities [118]. Applying DLTS pulses of opposite polarity (from accumulation to inversion) allows estimating the thermal generation times of bulk and surface centers [119]. DLTS techniques are capable in determining the trap properties in terms of relaxation mechanism and the defect profiling, the information valuable to study defects introduced by ion beams and ionizing radiation [100]. Naturally, characterization approaches are purpose-specific and can be based on several experimental techniques to identify a particular defect or study its energetics and kinetics. For example, commonly used techniques for studying the electrical- and optical characteristics of point defects such as DLTS and photoluminescence are sensitive to the defect states within the bandgap but have to be complemented by ESR studies to obtain information on the atomic structure of a defect or a defect complex.

4.4. Photoinjection

The methods based on photoinjection of charge carriers in metal-semiconductor barrier structures are sensitive to local nonuniformities in semiconductor interfaces because charge in a semiconductor induces an equal charge in the electrodes giving rise to electric fields at the interfaces, with a consequent field-effect modulation of the barrier heights (for the all-encompassing review on the subject of internal photoemission spectroscopy (IPE) methods one can refer to the book [120]). The early application of scanning internal photoemission to map sodium contamination at the Si/SiO_2 interface has been reported in work [121]. The IPE and trap photodepopulation methods were applied to reveal electron traps in Na^+ and Al^+ implanted SiO_2 [122]. At present, this technique has been revived to study ion beam induced charge nonuniformities in GaN and SiC [123].

Experimentally, the charge injected into an oxide, i.e., the current over the time of injection should remain unchanged by the method used for the charge detection. The trapped charge density is determined sensing the electric field created by the trapped charge. The electric field created by the charge of trapped carriers can be also observed in variations of the surface band bending of a semiconductor, i.e., a semiconductor space-charge layer serves as the field-sensing element. The band bending as a function of electric field can be extracted from capacitance-voltage measurements and the additional contribution of trapped charge to the field can be determined as a voltage shift of a CV curve. In MOSFETs, the trapped charge can be monitored as a function of the threshold voltage. This technique senses the charge carrier density in the inversion channel to monitor the electric field at the SI interface. Alternatively, the electric field induced by the trapped charge can be monitored by the Kelvin probe or photovoltage. In the latter case, the light intensity should be sufficient to set the flat band conditions at the semiconductor surface.

The experimental studies of the trapped charge in ion-implanted insulators are numerous with several examples represented in Refs. [124–131]. The interfacial defect densities modified by ion implantation have been studied combining IPE and *ac* conductance spectroscopy methods on nitrogen implanted SiC/SiO_2interfaces [132]. IPE reveals that nitrogen incorporates in carbon clusters at the SiC/SiO_2 interface that causes a shift of the electron levels to higher binding energies. Inferring the Schottky barrier height from the IPE spectra, it has been shown that ion implantation of sulfur in the NiSi/Si barrier does not induce changes in the barrier height, but increases doping in silicon [133]. The silicide/Si barrier modification by intentional dopant segregation has been verified in work [134].

4.5. Slow interface states as a special case of study

Defects generated by ionizing radiation and/or electric field, as well as the defects in undamaged devices, are considered to be spatially distributed across the SI interface and can be classified accordingly to the spatial location as the oxide-related traps and the interface traps. In respect to the latter, it is generally accepted that the interface traps are rapidly communicating with the silicon conduction or valence bands. The defects within the oxide interlayer also can exchange charge with silicon as has been revealed by the noise measurements [135]. Combining *ac* admittance spectroscopy and the noise measurements, it has been established that the fast interface states at the Si/SiO_2 interface, likely associated with the dangling bond defects, contribute to the loss peak in conductance measurements [136]. The defect states residing in an oxide layer are responsible for 1/f noise and random telegraph noise. These trapping centers in the oxide contribute to the conductance plateau at low frequency in *ac* conductance spectra (cf. **Figure 2** of Ref. [137]). A separable contribution of the oxide-related traps has been revealed employing measurements of subthreshold current [138] and the charge-pumping technique [139] to MOSFETs and CV measurements taken on the gate-controlled diode [140]. The latter technique is applicable for characterization of the interface traps in MOS devices composed on wide band gap semiconductors, because it allows supplying minority carriers in an amount sufficient to compensate for the low thermal generation rates of the minority carriers. An alternative method of providing minority carriers to invert a semiconductor surface is a controlled deposition of surface charges onto an insulator surface from corona discharging in air as it has been proposed in Ref. [141]. In this work, a surface charge has been deposited on SiO_2 and high-k dielectrics to overcompensate the carrier leakage current in silicon MOS capacitors and enable extraction of $D_{it}(E)$ profiles following the Berglund formalism. There are several advantages of the inverting semiconductor surfaces by employing noninvasive electrostatic charging of an insulator surface in a MOS structure: (i) The method does not involve fabrication of a transistor or a gate-controlled diode. (ii) The Berglund analysis can be used to reliably estimate $D_{it}(E)$ over the major part of a semiconductor band gap (for Si, from 0.2 to 0.9 eV above the valence band edge) using just MOS capacitors of one type of semiconductor conductivity. (iii) The method may employ CV measurements at mid-kHz frequency range allowing investigation of samples, which experience relatively high leakage current.

The sub-division of the interface trap responses into slow and fast on the basis of their characteristic time constants is important in research on the irradiation-induced damage in MOS devices. The interface state generation under irradiation or high electric field stress can involve electron-hole recombination in a gate insulator as proposed by Lai [142], the hole trapping according to the "hydrogen model" by Griscom [60], or generation of dangling bond defects in the oxide. Experimentally, it has been shown that both the fast and slow interface states can be generated upon oxide damage by high electric field or irradiation [143]. The mechanisms operative in the interface trap built-up upon irradiation or electric field stress are governed by hydrogen impurity, interfacial strain preexisting in thin insulating films on semiconductors, and experimental conditions used to impose damage on MOS devices.

Author details

Yanina G. Fedorenko

Address all correspondence to: janina.fedorenko@gmail.com

Stephenson Institute for Renewable Energy and Department of Physics, School of Physical Sciences, Chadwick Building, University of Liverpool, Liverpool, UK

References

[1] E. Rimini. Ion Implantation: Basics to Device Fabrication. Boston: Kluwer Academic Publishers; 1995. 393 p. DOI: 10.1007/978-1-4615-2259-1

[2] A. Dupasquier, A. P. Mills, editors. Positron Spectroscopy of Solids. Amsterdam: IOS; 1995. 805 p. DOI: 978-1-61499-211-0

[3] R. Krause-Rehberg, H. S. Leipner. Positron Annihilation in Semiconductors. Defect Studies. Berlin: Springer; 1999. XV, 383 p. DOI: 978-3-540-64371-5

[4] R. W. Siegell 1980. Positron annihilation spectroscopy. Annu. Rev. Sci. 1980;**10**:393–425. DOI: 10.1146/annurev.ms.10.080180.002141

[5] B. N. Ganguly. Positron annihilation spectroscopy: a prelude to modern aspects. Defect Diffus Forum. 2012; 331:7–21. DOI: 10.4028/www.scientific.net/DDF.331.7

[6] M. Alatalo, B. Barbiellini, M. Hakala, H. Kauppinen, T. Korhonen, M. J. Puska, K. Saarinen, P. Hautojärvi, R. M. Nieminen. Theoretical and experimental study of positron annihilation with core electrons in solids. Phys. Rev. B. 1996;54:2397. DOI: 10.1103/PhysRevB.54.2397

[7] F. Tuomisto, I. Makkonen. Defect identification in semiconductors with positron annihilation: experiment and theory. Rev. Mod. Phys. 2013;**85**(4):0034–6861. DOI: 10.1103/revmodphys.85.1583

[8] K.G. Lynn, A.N. Goland. Observation of high momentum tails of positron-annihilation lineshapes. Sol. State Commun. 1976;**18(11)**:1549–1552. DOI: 10.1016/0038-1098(76)90390-2

[9] K.G. Lynn, J.R. MacDonald, R.A. Boie, L.C. Feldman, J.D. Gabbe, M.F. Robbins, E. Bonderup, J. Golovchenko. Positron-annihilation momentum profiles in aluminum: core contribution and the independent-particle model. Phys. Rev. Lett. 1977;**38**:241. DOI: 10.1103/PhysRevLett.38.241

[10] A. P. Mills. Surface analysis and atomic physics with slow positron beams. Science. 1982;**218**:335–340. DOI: 10.1126/science.218.4570.335

[11] P.G. Coleman. Positron beams: the journey from fundamental physics to industrial application. Nuclear Instruments and Methods in Physics Research B. 2002;**192**:83–89. DOI: 10.1016/S0168-583X(02)00710-3

[12] A. Vehanen, J. Mäkinen. Thin films for slow positron generation. J. Appl. Phys. A. 1985;**36**:97–101. DOI: 10.1007/BF00620615

[13] R. Suzuki, T. Ohdaira, A. Uedono, Y. K. Cho, S. Yoshida, Y. Ishida, T. Ohshima, H. Itoh, M. Chiwaki, T. Mikado. Investigation of positron moderator materials for electron-linac-based slow positron beamlines. Jpn. J. Appl. Phys. 1998;**37**:4636. DOI: 10.1143/JJAP.37.4636

[14] P. Coleman. The generation and transport of positron beams. In: P. Coleman, editor. Positron Beams and Their Applications. World Scientific; Singapore, 2000. pp. 11–40. DOI: 10.1142/9789812817754_0002

[15] R. M. Nieminen. The fate of slow positrons in condensed matter. In: P. Coleman, editor. Positron Beams and Their Applications. World Scientific; Singapore, 2000. pp. 97–128. DOI: 10.1142/9789812817754_0002

[16] M. Charlton, J. W. Humberston. Positron Physics. Cambridge: Cambridge University Press; 2001. 454 p. ISBN:0521415500

[17] A. Weiss, R. Mayer, M. Jibaly, C. Lei, D. Meh, K. G. Lynn. Auger-electron emission resulting from the annihilation of core electrons with low-energy positrons. Phys. Rev. Lett. 1988;**61**:2245. DOI: 10.1103/PhysRevLett.61.2245

[18] G. Chiarotti, P. Chiaradia, editors. Physics of Solid Surfaces. Berlin: Springer; 2015. XXII, 750 p. DOI: 10.1007/978-3-662-47736-6

[19] C. Hugenschmidt. Positrons in surface physics. Surf. Sci. Rep. 2016;**71**:547–594. DOI: 10.1016/j.surfrep.2016.09.002

[20] P. G. Coleman, A. P. Knights, R. M. Gwilliam. Diagnostic measurement of ion implantation dose and uniformity with a laboratory-based positron probe. J. Appl. Phys. 1999;**86**:5988. DOI: 10.1063/1.371644

[21] C. Szeles, P. Asoka-Kumar, and K. G. Lynn H.-J. Gossmann, F. C. Unterwald, T. Boone. Defect distribution in low-temperature molecular beam epitaxy grown Si/Si(100), improved depth profiling with monoenergetic positrons. Appl. Phys. Lett. 1995;**66**: 2855. DOI: 10.1063/1.113452

[22] A. Uedono, S. Tanigawa, J. Sugiura, M. Ogasawara. A study of vacancy-type defects in B^+-implanted SiO_2/Si by a slow positron beam. Jpn. J. Appl. Phys. 1989;**28**:1293. DOI: 10.1143/JJAP.28.1293

[23] A. Uedono, S. Tanigawa, T. Ohshima, H. Itoh, M. Yoshikawa. Crystallization of an amorphous layer in P^+-implanted 6H-SiC studied by monoenergetic positron beams. J. Appl. Phys. 2000;**87**:4119. DOI: 10.1063/1.373039

[24] Y. Kobayashi, I. Kojima, S. Hishita, T. Suzuki, E. Asari, M. Kitajima. Damage-depth profiling of an ion-irradiated polymer by monoenergetic positron beams. Phys. Rev. B. 1995;**52**:823. DOI: 10.1103/PhysRevB.52.823

[25] E.H. Lee, G.R. Rao, L.K. Mansur. LET effect on cross-linking and scission mechanisms of PMMA during irradiation. Radiat. Phys. Chem. 1993;**55**:293. DOI: 10.1016/S0969-806X(99)00184-X

[26] S. Bangsaruntip, K. Balakrishnan, S.-L. Cheng, J. Chang, M. Brink, I. Lauer, R. L. Bruce, S. U. Engelmann, A. Pyzyna, G. M. Cohen, L. M. Gignac, C. M. Breslin, J. S. Newbury, D. P. Klaus, A. Majumdar, J. W. Sleight, M. A. Guillorn. Density scaling with gate-all-around silicon nanowire MOSFETs for the 10 nm node and beyond. Electron Devices Meeting (IEDM), 2013 IEEE International. 2013;DOI: 10.1109/IEDM.2013.6724667

[27] G. Hellings, K. De Meyer. High Mobility and Quantum Well Transistors. Dordrecht: Springer; 2013. 140 p. DOI: 10.1007/978-94-007-6340-1

[28] T. Schenkel, A. Persaud, S. J. Park, J. Nilsson, J. Bokor, J. A. Liddle, R. Keller, D. H. Schneider, D. W. Cheng, D. E. Humphries. Solid state quantum computer development in silicon with single ion implantation. J. Appl. Phys. 2003;**94**:7017. DOI: 10.1063/1.1622109

[29] P. Neumann, R. Kolesov, B. Naydenov, J. Beck, F. Rempp, M. Steiner, V. Jacques, G. Balasubramanian, M. L. Markham, D. J. Twitchen, S. Pezzagna, J. Meijer, J. Twamley, F. Jelezko, and J. Wrachtrup. Quantum register based on coupled electron spins in a room-temperature solid. Nat. Phys. 2010;**6**:249 –253. DOI: 10.1038/nphys1536

[30] D. L. Young, W. Nemeth, V. LaSalvia, M. R. Page, S. Theingi, J. Aguiar, B. G. Lee, P. Stradins. Low-cost plasma immersion ion implantation doping for interdigitated back passivated contact (IBPC) solar cells, Solar Energy Mater. Solar Cells. 2016;**158**:68–76. DOI: 10.1016/j.solmat.2016.05.044

[31] F. Kiefer, R. Peibst, T. Ohrdes, T. Dullweber, J. Krügener, H. Jörg Osten, C. Schöllhorn, A. Grohe, R. Brendel. Influence of the boron emitter profile on V_{OC} and J_{SC} losses in fully ion implanted n-type PERT solar cells, Stat. Sol. (a). 2015;**(212)**:291–297. DOI: 10.1002/pssa.201431118

[32] J. Duchaine, F. Milési, R. Coquand, S. Barraud, S. Reboh, F. Gonzatti, F. Mazen, F. Torregrosa. Plasma immersion ion implantation for sub-22 nm node devices: FD-SOI and Tri-Gate. AIP Conference Proceedings. 2012;1496:71–74. DOI: 10.1063/1.4766492

[33] S. Qin, Y. Jeff Hu, A. McTeer. PLAD (plasma doping) on 22nm technology node and beyond—evolutionary and/or revolutionary, junction technology (IWJT). (2012 12th International Workshop). DOI: 10.1109/IWJT.2012.6212800

[34] E. Napolitani, G. Impellizzeri. Ion implantation defects and shallow junctions in Si and Ge. Semiconduct. Semimet. 2015;**91**:93–122. DOI: 10.1016/bs.semsem.2015.01.001

[35] A. Uedono, T. Naito, T. Otsuka, K. Ito, K. Shiraishi, K. Yamabe, S. Miyazaki1, H. Watanabe, N. Umezawa, T. Chikyow. Characterization of metal/high-k structures using monoenergetic positron beams. Jpn. J. Appl. Phys. 2007;**46**:3214. DOI: 10.1143/JJAP.46.3214

[36] A. Uedonoa, R. Hasunumaa, K. Shiraishia, K. Yamabea, S. Inumiyab, Y. Akasakac, S. Kamiyamad, T. Matsukid, T. Aoyamad, Y. Narad, S. Miyazakie, H. Watanabef, N. Umezawag, T. Chikyowh, S. Ishibashii, T. Ohdairai, R. Suzukii, K. Yamadaj. Vacancy-type defects in MOSFETs with high-k gate dielectrics probed by monoenergetic positron beams. ECS Trans. 2007;**11**(4):81–90. DOI: 10.1149/1.2779550

[37] T. Matsuki, T. Watanabe, T. Miura, N. Mise, T. Eimori, Y. Nara, Y. Ohji, A. Uedono, K. Yamada. Impact of high temperature annealing on traps in physical-vapor-deposited-TiN/SiO_2/Si analyzed by positron annihilation. Jpn. J. Appl. Phys. 2007;**46**:L1219. DOI: 10.1143/JJAP.46.L1219

[38] A. Uedonoa, K. Ikeuchi, T. Otsuka, K. Ito, K. Yamabe M. Kohno, T. Moriya, N. Okumura, T. Nakanishi, T. Arikado T. Ohdaira, R. Suzuki. Open volumes in SiN films for strained Si transistors probed using monoenergetic positron beams. Appl. Phys. Lett. 2006;**88**:252107. DOI: 10.1063/1.2216901

[39] C.F. Hsieh, C.W. Chen, C.H. Chen, M.H. Liao. Optimized $Si_{0.5}Ge_{0.5}$/Si interface quality by the process of low energy hydrogen plasma cleaning and investigation by positron annihilation spectroscopy. Proc. Eng. 2014;**79**:328–332. DOI: 10.1016/j.proeng.2014.06.351

[40] S. Garaj, W. Hubbard, J.A. Golovchenko. Graphene synthesis by ion implantation. Appl. Phys. Lett. 2010;**97**:183103. DOI: 10.1063/1.3507287

[41] H. Tsai, C. Hsiao, C. Chen, H. Ouyang, J. Liang. Synthesis of nonepitaxial multilayer silicene assisted by ion implantation. Nanoscale. 2016;**8**:9488–9492. DOI: 10.1039/C6NR02274J

[42] M. Hori, M. Uematsu, A. Fujiwara, Y. Ono. Electrical activation and electron spin resonance measurements of arsenic implanted in silicon. Appl. Phys. Lett. 2015;**106**:142105. DOI: 10.1063/1.4917295

[43] W. L. Ng, M. A. Lourenço, R. M. Gwilliam, S. Ledain, G. Shao, K. P. Homewood. An efficient room-temperature silicon-based light-emitting diode. Nature. 2001;**410**:192–194. DOI: 10.1038/35065571

[44] B.J. Mrstik, H.L. Hughes, P.J. McMarr, R.K. Lawrence, D.I. Ma, I.P. Isaacson, R.A. Walker. Hole and electron trapping in ion implanted thermal oxides and SIMOX. IEEE Transactions on Nuclear Science. 2000;**47**(6):2189–2195. DOI: 10.1109/23.903752

[45] R. A. B. Devine. The structure of SiO_2, its defects and radiation hardness. IEEE Trans. Nucl. Sci. 1994;**41**:452–459. DOI: 10.1109/23.299784

[46] M. Fanciulli, editor. Electron Spin Resonance and Related Phenomena in Low-Dimensional Structures. Berlin, Heidelberg: Springer; 2009. 276 p. DOI: 10.1007/978-3-540-79365-6

[47] K. Marumoto, S. Kuroda, T. Takenobu, Y. Iwasa. Spatial extent of wave functions of gate-induced hole carriers in pentacene field-effect devices as investigated by electron spin resonance. Phys. Rev. Lett. 2006; **97**:256603. DOI: 10.1103/PhysRevLett.97.256603

[48] A. Stesmans and V.V. Afanas'ev. Electron spin resonance features of interface defects in thermal (100)Si/SiO_2. J. Appl. Phys. 1998;**83**:2449. DOI: 10.1063/1.367005

[49] A. Stesmans. Passivation of P_{b0} and P_{b1} interface defects in thermal (100)Si/SiO_2 with molecular hydrogen. Appl. Phys. Lett. 1996;**68**:2076. DOI: 10.1063/1.116308

[50] D. Hiller, M. Jivanescu, A. Stesmans, M. Zacharias. P_{b0} centers at the Si-nanocrystal/SiO_2 interface as the dominant photoluminescence quenching defect. J. Appl. Phys. 2010;**107**:084309. DOI: 10.1063/1.3388176

[51] H. M. Assenheim, editor. Electron Spin Resonance in Semiconductors. 2nd ed. New York: Plenum Press; 1967. 152 p. DOI: 10.1007/978-1-4684-1737-1

[52] J. E. Wertz, J. R. Bolton. Electron Spin Resonance: Elementary Theory and Practical Applications. 2nd ed. New York, London: Chapman and Hall; 1986. 497 p. DOI: 10.1007/978-94-009-4075-8

[53] A. Stesmans, V. V. Afanas′ev. Electron spin resonance of interfaces and nanolayers in semiconductors. In: Characterization of Semiconductor Heterosctructures and Nanosctructures. 2nd ed. 2013. Elsevier, the Netherlands, the UK, pp. 685–752. DOI:10.1016/B978-0-444-59551-5.00016-9

[54] A. Stesmans, B. Nouwen, and V.V. Afanas'ev. Structural degradation of thermal SiO_2 on Si by high-temperature annealing: defect generation. Phys. Rev. B. 2002;**66**:045307. DOI: 10.1103/PhysRevB.66.045307

[55] Y. Nishi. Study of silicon-silicon dioxide structure by electron spin resonance. Jpn. J. Appl. Phys. 1971; **110**: 52. DOI: 10.1143/JJAP.10.52

[56] A. Stesmans. Electron spin resonance of $[1\,\bar{1}1]$, $[\bar{1}\,1\,1]$, and $[1 1\,\bar{1}]$ oriented dangling orbital P_{b0} defects at the (111)Si/SiO_2 interface. Appl. Phys. Lett. 1986;**48**:972. DOI: 10.1063/1.96627

[57] G. Lucovsky, Y. Wu, H. Niimi, V. Misra, J. C. Phillips. Bonding constraint-induced defect formation at Si-dielectric interfaces and internal interfaces in dual-layer gate dielectrics. J. Vac. Sci. Technol. 1999;**B17**:1806. DOI: 10.1116/1.590831

[58] K. L. Brower. Strain broadening of the dangling-bond resonance at the (111)Si-SiO_2 interface. Phys. Rev. B. 1986; **33:**4471. DOI: 10.1103/PhysRevB.33.4471

[59] A. Stesmans, J. Braet. In: J.J. Simonne, J. Buxo, editors. Insulating Films on Semiconductors. North-Holland ed. Amsterdam; 1986. p. 25.

[60] D. L. Griscom. Self-trapped holes in pure silica glass: a history of their discovery and characterisation and an example of their critical significance to industry. J. Non-Cryst. Sol. 2006;**352**:2601–2617. DOI: 10.1016/j.jnoncrysol.2006.03.033

[61] M. Kaviani, J. Strand, V. V. Afanas'ev, A. L. Shluger. Deep electron and hole polarons and bipolarons in amorphous oxide. Phys. Rev. B. 2016;**94**:020103(R). DOI: 10.1103/PhysRevB.94.020103

[62] A. Stesmans, F. Scheerlinck. Parameters controlling the generation of natural intrinsic *EX* defects in thermal SiO_2 on Si. J. Non-Cryst. Sol. 1995;**187**:119–123. DOI: 10.1016/0022-3093(95)00123-9

[63] M. Stapelbroek, D.L. Griscom, E.J. Friebele, G.H. Sigel Jr. Oxygen-associated trapped-hole centers in high-purity fused silicas. J. Non-Cryst. Sol. 1979;**32**:313–326. DOI: 10.1016/0022-3093(79)90079-6

[64] E. J. Friebele, D. L. Griscom, M. Stapelbroek, R. A. Weeks. Fundamental defect centers in glass: the peroxy radical in irradiated, high-purity, fused silica. Phys. Rev. Lett. 1979;**42**:1346. DOI: 10.1103/PhysRevLett.42.1346

[65] W. L. Warren, E. H. Poindexter, M. Offenberg, W. Muller-Warmuth. Paramagnetic point defects in amorphous silicon dioxide and amorphous silicon nitride thin films. Part 1. a-SiO_2. J. Electrochem. Soc. 1992;**139**:872. DOI: 10.1103/PhysRevLett.42.1346

[66] T. Uchino, M. Takahashi, T. Yoko. E′ centers in amorphous SiO_2 revisited: a new look at an old problem. Phys. Rev. Lett. 2001;**86**:5522. DOI: 10.1103/PhysRevLett.86.5522

[67] Al-M. El-Sayed, M. B. Watkins, T. Grasser, V. V. Afanas'ev, A. L. Shluger. Hydrogen induced rupture of Si–O bonds in amorphous silicon dioxide. Phys. Rev. Lett. 2015;**114**:115503. DOI: 10.1103/PhysRevLett.86.5522

[68] V. Zekeriya, T. P. Ma. Interface traps generated by internal photoemission in Al-SiO_2-Si structures. Appl. Phys. Lett. 1983;**43**:95. DOI: 10.1063/1.94135

[69] Y. Nishi, K. Tanaka, A. Ohwada. Study of silicon-silicon dioxide structure by electron spin resonance II. Jpn. J. Appl. Phys. 1972;**11**:85. DOI: 10.1103/PhysRevLett.86.5522

[70] E. H. Poindexter, G. J. Gerardi, M.-E. Rueckel, P. J. Caplan, N. M. Johnson, D. K. Biegelsen. Electronic traps and P_b centers at the Si/SiO_2 interface: band-gap energy distribution. J. Appl. Phys. 1984;**56**(2844). DOI: 10.1063/1.333819

[71] H. G. Grimmeiss, W. R. Buchwald, E. H. Poindexter, P. J. Caplan, M. Harmatz, G. J. Gerardi, D. J. Keeble, N. M. Johnson. Optical and electrical studies of interface traps in the Si/SiO_2 system by modified junction space-charge techniques. Phys. Rev. B. 1989;**39**:5175. DOI: 10.1103/PhysRevB.39.5175

[72] N. M. Johnson, W. Shan, P. Y. Yu. Pressure dependence of the P_b center at the (111)Si/SiO_2 interface. Phys. Rev. B. 1989;**39**:3431(R). DOI: 10.1103/PhysRevB.39.3431

[73] G. J. Gerardi, E. H. Poindexter, P. J. Caplan, N. M. Johnson. Interface traps and P_b centers in oxidized (100) silicon wafers. Appl. Phys. Lett. 1986;**49**:348. DOI: 10.1063/1.97611

[74] A. Stesmans, V. V. Afanas'ev. Electrical activity of interfacial paramagnetic defects in thermal (100)Si/SiO_2. Phys. Rev. B. 1998;**57**:10030. DOI: 10.1103/PhysRevB.39.3431

[75] V. V. Afanas'ev, A. Stesmans. Charge state of paramagnetic E' centre in thermal SiO_2 layers. J. Phys.: Condens. Matter. 2000;**12**:2285–2290. DOI: 10.1088/0953-8984/12/10/312

[76] A. Stesmans, K. Clémer, V. V. Afanas'ev. P-associated defects in the high-k insulators HfO_2 and ZrO_2 revealed by electron spin resonance. Phys. Rev. B. 2008;**77**:125341. DOI: 10.1103/PhysRevB.77.125341

[77] M. Jivanescu, A. Romanyuk, A. Stesmans. Influence of *in situ* applied ultrasound during Si^+ implantation in SiO_2 on paramagnetic defect generation. J. Appl. Phys. 2010;**107**:114307. DOI: 10.1063/1.3369041

[78] D. Tsvetanova, R. Vos, G. Vereecke, T. N. Parac-Vogt, F. Clemente, K. Vanstreels, D. Radisic, T. Conard, A. Franquet, M. Jivanescu, D. A. P. Nguyen, A. Stesmans, B. Brijs, P. Mertens, M. M. Heyns. Degradation of 248 nm deep UV photoresist by ion implantation. J. Electrochem. Soc. 2011;**158**:H785–H794. DOI: 10.1149/1.3597176

[79] M. S. Dresselhaus, R. Kalish. Ion Implantation in Diamond, Graphite and Related Materials. Berlin, Heidelberg: Springer; 1992. 202 p. DOI: 10.1007/978-3-642-77171-2

[80] T. Schenkel, J. A. Liddle, A. Persaud, A. M. Tyryshkin, S. A. Lyon, R. de Sousa and K. B. Whaley, J. Bokor, J. Shangkuan, I. Chakarov. Electrical activation and electron spin coherence of ultralow dose antimony implants in silicon. Appl. Phys. Lett. 2006;**88**:112101. DOI: 10.1063/1.2182068

[81] D. M. Toyli, C. D. Weis, G. D. Fuchs, T. Schenkel, D. D. Awschalom. Chip-scale nanofabrication of single spins and spin arrays in diamond. Nano Lett. 2010;**10**:3168–3172. DOI: 10.1021/nl102066q

[82] D. K. Schroder. Semiconductor Material and Device Characterization. 3rd ed. Hokoben, U.S.A.: John Wiley & Sons, 2006. 800 p. DOI: 10.1002/0471749095

[83] C.N. Berglund. Surface states at steam-grown silicon-silicon dioxide interfaces. IEEE Trans. Electron Dev. 1966;**3**:701–705. DOI: 10.1109/T-ED.1966.15827

[84] T. Sakurai, T. Sugano. Theory of continuously distributed trap states at Si-SiO_2 interfaces. J. Appl. Phys.1981;**52**:2889. DOI: 10.1063/1.329023

[85] P. V. Gray, D. M. Brown. Density of SiO_2–Si interface states. Appl. Phys. Lett. 1966;**8**:31. DOI: 10.1063/1.1754468

[86] L. M. Terman. An investigation of surface states at a silicon/silicon oxide interface employing metal-oxide-silicon diodes. Sol. State Electron. 1962;**5**:285-299. DOI: 10.1016/0038-1101(62)90111-9

[87] E. M. Vogel, A. M. Sonnet, C. L. Hinkle. Characterization of electrically active interfacial defects in high-κ gate dielectrics. ECS Trans. 2007;**11(4)**:393–406. DOI: 10.1149/1.2779576

[88] D. P. Kennedy, P. C. Murley, W. Kleinfelder. On the measurement of impurity atom distributions in silicon by the differential capacitance technique. IBM J. Res. Dev. 1968;**12**:399–409. DOI: 10.1147/rd.125.0399

[89] P. Blood. Capacitance-voltage profiling and the characterisation of III-V semiconductors using electrolyte barriers. Semicond. Sci. Technol. **1**:7. DOI: 10.1088/0268-1242/1/1/002

[90] A. Koeder, S. Frank, W. Schoch, V. Avrutin, W. Limmer, K. Thonke, R. Sauer, A. Waag, M. Krieger, K. Zuern, P. Ziemann, S. Brotzmann, H. Bracht. Curie temperature and carrier concentration gradients in epitaxy-grown $Ga_{1-x}Mn_xAsGa_{1-x}Mn_xAs$ layers. Appl. Phys. Lett. 2003;**82** :3287. DOI: 10.1063/1.1573369

[91] M. Green. Electrochemistry of the semiconductor-electrolyte electrode. I. The electrical double layer. J. Chem. Phys. 1959;**31**:200–203. DOI: 10.1063/1.1730295

[92] A. J. Nozik, R. Memming. Physical chemistry of semiconductor-liquid interfaces. J. Phys. Chem. 1996;**100 (31)**:13061–13078. DOI: 10.1021/jp953720

[93] R. Rios, N. D. Arora. Determination of ultra-thin gate oxide thicknesses for CMOS structures using quantum effects. Electron Devices Meeting. IEDM '94. Technical Digest, 25.6.1. 1994;DOI: 10.1109/IEDM.1994.383335

[94] F. Rana, S. Tiwari, D. A. Buchanan. Self-consistent modeling of accumulation layers and tunneling currents through very thin oxides. Appl. Phys. Lett. 1996;**69**:1104-1106. DOI: 10.1063/1.117072

[95] S. Takagi, A. Toriumi. Quantitative understanding of inversion-layer capacitance in Si MOSFET's. IEEE Trans. Electron Dev. 1995;**42**:2125–2130. DOI: 10.1109/16.477770

[96] M. Sze, Physics of Semiconductor Devices. Wiley ed. New York: 1969. 432-436 p.

[97] V. V. Afanas'ev, J. M. M. de Nijs, P. Balk, A. Stesmans. Degradation of the thermal oxide of the $Si/SiO_2/Al$ system due to vacuum ultraviolet irradiation. J. Appl. Phys. 1995;**78**:6481. DOI: 10.1063/1.360534

[98] V. V. Afanas'ev, J. M. M. de Nijs, P. Balk. Elimination of hydrogen related instabilities in Si/SiO_2 structures by fluorine implantation. J. Appl. Phys. 1994;**76**:7990. DOI: 10.1063/1.357911

[99] P. De Wolf, R. Stephenson, T. Trenkler, T. Clarysse, T. Hantschel, W. Vandervorst. Status and review of two-dimensional carrier and dopant profiling using scanning probe microscopy. J. Vac. Sci. Technol. B. 2000;**18**:361. DOI: 10.1116/1.591198

[100] K. Yamasaki, M. Yoshida, T. Sugano. Deep level transient spectroscopy of bulk traps and interface states in Si MOS diodes. Jpn. J. Appl. Phys. 1979;18(113). DOI: 10.1143/JJAP.18.113

[101] J. F. Cordaro, Y. Shim, J. E. May, Bulk electron traps in zinc oxide varistors. J. Appl. Phys. 1986;**60**(4186). DOI: 10.1063/1.337504

[102] M. J. Uren, K. M. Brunson, A. M. Hodge. Separation of two distinct fast interface state contributions at the (100)Si/SiO_2 interface using the conductance technique. Appl. Phys. Lett. 1992;**60**:624. DOI: 10.1063/1.107461

[103] G. Pensl, F. Ciobanu, T. Frank, D. Kirmse, M. Krieger, S. Reshanov, F. Schmid, M. Weidner, T. Ohshima, H. Itoh, W.J. Choyke. Defect-engineering in SiC by ion implantation and electron irradiation. Microelectron. Eng. 2006;**83**: 146–149. DOI: 10.1016/j.mee.2005.10.040

[104] E. H. Nicollian, J. R. Brews. MOS (Metal Oxide Semiconductor) Physics and Technology. U.S.A.: John Wiley & Sons, 2002. 978 p. ISBN: 978-0-471-43079-7

[105] E. H. Nicollian, A. Goetzberger. The Si-SiO_2 interface—electrical properties as determined by the metal-insulator-silicon conductance technique. Bell Labs Techn. J. 1967; **XLVI**:1055-1133. DOI: 10.1002/j.1538-7305.1967.tb01727.x

[106] H. Preier. Contributions of surface states to MOS impedance. Appl. Phys. Lett. 1967;**10**:361–363. DOI: 10.1063/1.1728213

[107] S. Collins, M. J. Kirton, M. J. Uren. Asymmetric conductance peaks observed in silicon metal-oxide-semiconductor capacitors. Appl. Phys. Lett. 1990;**57** :372 . DOI: 10.1063/1.103695

[108] E.M. Vogel, W.K. Henson, C.A. Richter, J.S. Suehle. Limitations of conductance to the measurement of the interface state density of MOS capacitors with tunneling gate dielectrics. IEEE Trans. Electron Dev. 2000;**47**:601–608. DOI: 10.1109/16.824736

[109] L. Autran, F. Seigneur, C. Plossu, B. Balland. Characterization of Si-SiO_2 interface states: comparison between different charge pumping and capacitance techniques. J. Appl. Phys. 1993;**74**:3932. DOI: 10.1063/1.354493

[110] T. Aichinger, M. Nelhiebel. Characterization of MOSFET interface states using the charge pumping technique. In: T. Grasser, editor. Hot Carrier Degradation in Semiconductor Devices. Springer; Switzerland, 2015. pp. 231-255. DOI: 10.1007/978-3-319-08994-2_8

[111] P. Masson, J.-L. Autran, M. Houssa, X. Garros, C. Leroux. Frequency characterization and modeling of interface traps in $HfSi_xO_y/HfO_2$ gate dielectric stack from a capacitance point-of-view. Appl. Phys. Lett. 2002;**81**:3392–3394. DOI: 10.1063/1.1518561

[112] O. Engström, B. Raeissi, J. Piscator. Vibronic nature of hafnium oxide/silicon interface states investigated by capacitance frequency spectroscopy. J. Appl. Phys. 2008;**103**: 104101. DOI: 10.1063/1.2921795

[113] Y. G. Fedorenko, V. V. Afanas'ev, A. Stesmans. Impact of Al incorporation in hafnia on interface states in (100)Si/$HfAl_xO_y$. Microelectron. Eng. 2005;**80**:66–69. DOI: 10.1016/j.mee.2005.04.045

[114] L. Dobaczewski, A. R. Peaker, K. Bonde Nielsen. Laplace-Transform deep-level spectroscopy: the technique and its applications to the study of point defects in semiconductors. J. Appl. Phys. 2004;**96**:4689. DOI: 10.1063/1.1794897

[115] N. M. Johnson. Energy resolved DLTS measurement of interface states in MIS structures. Appl. Phys. Lett. 1979;**34**:802. DOI: 10.1063/1.90650

[116] T. Katsube, K. Kakimoto, T. Ikoma. Temperature and energy dependences of capture cross sections at surface states in Si metal-oxide-semiconductor diodes measured by deep level transient spectroscopy. J. Appl. Phys. 1981;**52**:3504. DOI: 10.1063/1.329128

[117] T. Hashizume, H. Hasegawa, R. Riemenschneider, H. L. Hartnagel. Process-induced defects in InP caused by chemical vapor deposition of surface passivation dielectrics. Jpn. J. Appl. Phys. 1994;**33**:727. DOI: 10.1143/JJAP.33.727

[118] A. G. Aberle, S. Glunz, W. Warta. Impact of illumination level and oxide parameters on Shockley-Read-Hall recombination at the Si-SiO_2 interface. J. Appl. Phys. 1992;**71**:4422. DOI: 10.1063/1.350782

[119] A. E. Khorasani, D. K. Schroder, T. L. Alford. A fast technique to screen carrier generation lifetime using DLTS on MOS capacitors. IEEE Trans. Electron Dev. 2014;**61**:3282–3288. DOI: 10.1109/TED.2014.2337898

[120] V. V. Afanas'ev. Internal Photoemission Spectroscopy. Principles and Applications. Elsevier, UK and the Netherlands, ed. 2008. 312 p.

[121] T. H. DiStefano. Barrier inhomogeneities on a Si–SiO_2 interface by scanning internal photoemission. Appl. Phys. Lett. 1971;**19**:280. DOI: 10.1063/1.1653918

[122] E. Harari, B. S. H. Royce. Oxide charge trapping induced by ion implantation in SiO_2. IEEE Trans. Nucl. Sci. 1973;**20**:288–292. DOI: 10.1109/TNS.1973.4327409

[123] S. Murase, T. Mishima, T. Nakamura, K. Shiojima. Mapping of ion-implanted n-SiC Schottky contacts using scanning internal photoemission microscopy. Mater. Sci. Semicond. Process. In Press. DOI: 10.1016/j.mssp.2016.10.055

[124] D.M. Fleetwood, L.C. Riewe, J.R. Schwank, S.C. Witczak, R.D. Schrimpf. Radiation effects at low electric fields in thermal, SIMOX, and bipolar-base oxides. IEEE Trans. Nucl. Sci. 1996;**43**:2537-2546. DOI: 10.1109/23.556834

[125] R.E. Stahlbush, H.L. Hughes, W.A. Krull. Reduction of charge trappings and electron tunneling in SIMOX by supplemental implantation of oxygen. IEEE Trans. Nucl. Sci. 1993;**40**:1740–1747. DOI: 10.1109/23.273484

[126] V. V. Afanas'ev, G. A. Brown, H. L. Hughes, S. T. Liu, A. G. Revesz. Conducting and charge-trapping defects in buried oxide layers of SIMOX structures. J. Electrochem. Soc. 1996;**143**:347–352. DOI: 10.1149/1.1836434

[127] G. Zhang, Z. Liu, N. Li, Z. Zhen, G. Li. Influence of fluorine on radiation-induced charge trapping in the SIMOX buried oxides. Solid-State and Integrated Circuits Technology Proceedings. 7th International Conference on 18-21 Oct. 2004. 2004; DOI: 10.1109/ICSICT.2004.1436640

[128] H. Huang, Y. Yang Huang, J. Zheng, S. Wei, K. Tang, D. Bib, Z. Zhang. Hardening silicon-on-insulator nMOSFETs by multiple-step Si^+ implantation. Microelectron. Reliab. 2016;**57**:1–9. DOI: 10.1016/j.microrel.2015.12.015

[129] S. Cristoloveanu, M. Bawedin, I. Ionica. A review of electrical characterization techniques for ultrathin FDSOI materials and devices. Sol. State Electron. 2016;**117**:10–36. DOI: 10.1016/j.sse.2015.11.007

[130] K. Potter, K. Morgan, C. Shaw, P. Ashburn, W. Redman-White, C.H. De Groot. Total ionizing dose response of fluorine implanted silicon-on-insulator buried oxide. Microelectron. Reliab. 2014;**54**:2339–2343. DOI: 10.1016/j.microrel.2014.07.018

[131] Y. Li, W. M. Porter, C. Kshirsagar, I. Roth, Y. Su, M. A. Reynolds, B. J. Gerbi, S. J. Koester. Fully-depleted silicon-on-insulator devices for radiation dosimetry in cancer therapy. IEEE Trans. Nucl. Sci. 2014;**61**:3443–3450. DOI: 10.1109/TNS.2014.2365544

[132] F. Ciobanu, T. Frank, G. Pensl, V. Afanas'ev, S. Shamuilia, A. Schöner, T. Kimoto. Nitrogen implantation – an alternative technique to reduce traps at SiC/SiO_2-interfaces. Mater. Sci. Forum. 2006;**527-529**:991–994

[133] Chan, N. Y. Martinez, J. J. D. Fitzgerald, A. V. Walker, R. A. Chapman, D. Riley, A. Jain, C. L. Hinkle, E.M. Vogel. Extraction of correct Schottky barrier height of sulfur implanted NiSi/n-Si junctions: junction doping rather than barrier height lowering. Appl. Phys. Lett. 2011;**99**:012114. DOI: doi: 10.1063/1.3609874

[134] Z. Zhang, J. Atkin, M. Hopstaken, M. Hatzistergos, P. Ronsheim, E. Liniger, R. Laibowitz, P. M. Solomon. Probing the interface barriers of dopant-segregated silicide–Si diodes with internal photoemission. IEEE Trans. ED. 2012;**59**:2027–2032. DOI: 10.1109/TED.2012.2197399

[135] S. Christensson, I. Lundström, C. Svensson. Low frequency noise in MOS transistors—I. Theory. Sol. State Electron. 1968;**11**:797–812. DOI: 10.1016/0038-1101(68)90100-7

[136] M.J. Kirton, M.J. Uren. Noise in solid-state microstructures: a new perspective on individual defects, interface states and low-frequency ($1/f$) noise. Adv. Phys. 1989;**38:4**:367–468. DOI: 10.1080/00018738900101122

[137] M. J. Uren, S. Coilins, M. J. Kirton. Observation of "slow" states in conductance measurements on silicon metal-oxide-semiconductor capacitors. Appl. Phys. Lett. 1989;**54**: 1448. DOI: 10.1063/1.100693

[138] P. J. McWhorter, P. S. Winokur. Simple technique for separating the effects of interface traps and trapped oxide charge in metal-oxide-semiconductor transistors. Appl. Phys. Lett. 1986;**48**:133. DOI: 10.1063/1.96974

[139] R.E. Paulsen, R.R. Siergiej, M.L. French, M.H. White. Observation of near-interface oxide traps with the charge-pumping technique. IEEE Electron Dev. Lett. 1992;**13**:627–629. DOI: 10.1109/55.192866

[140] S. T. Sheppard, M. R. Melloch, J. A. Cooper. Characteristics of inversion-channel and buried-channel MOS devices in 6H-SiC. IEEE Trans. Electron Dev. 1994;**41**:1257–1264. DOI: 10.1109/16.293356

[141] Y. G. Fedorenko, L. Truong, V. V. Afanas'ev, A. Stesmans. Energy distribution of the (100) Si/HfO_2 interface states. Appl. Phys. Lett. 2004;**84**:4771. DOI: 10.1063/1.1758302

[142] S. K. Lai. Interface trap generation in silicon dioxide when electrons are captured by trapped holes. J. Appl. Phys. 1983;**54**:2540. DOI: 10.1063/1.332323

[143] D.M. Fleetwood. Effects of hydrogen transport and reactions on microelectronics radiation response and reliability. Microelectron. Reliab. 2002;**42**:523–541. DOI: 10.1016/S0026-2714(02)00019-7

[144] S. Szpala, P. Asoka-Kumar, B. Nielsen, J. P. Peng, S. Hayakawa, K. G. Lynn, H.-J. Gossmann. Defect identification using the core-electron contribution in Doppler-broadening spectroscopy of positron-annihilation radiation. Phys. Rev. B. 1996;**54**:4722–4731. DOI: 10.1103/PhysRevB.54.4722

[145] E. H. Poindexter, P. J. Caplan. Characterization of Si/SiO_2 interface defects by electron spin resonance. Progr. Surf. Sci. 1983;**14**:201–294. DOI: 10.1016/0079-6816(83)90006-0

Section 3

Device Fabrication

Chapter 5

Development of Optical Waveguides Through Multiple-Energy Ion Implantations

Heriberto Márquez Becerra, Gloria V. Vázquez,
Eder G. Lizárraga-Medina, Raúl Rangel-Rojo,
David Salazar and Alicia Oliver

Additional information is available at the end of the chapter

http://dx.doi.org/10.5772/67829

Abstract

In this chapter, we present information about the design, fabrication and characterization of optical waveguides obtained by using a protocol of multiple energy ion implantations. This protocol must provide an approach to produce optical waveguides with adequate features, such as dimensions, evanescent field and optical confinement. In general, optical waveguides can be improved by widening the optical barrier or waveguide core through multiple energy ion implantations. Design of optical waveguides must consider effects induced by the ion implantation process, such as modification of substrate density, polarizability and structure. Information will be presented about optical waveguides formed mainly in laser crystals (i.e., Nd:YAG, Nd:YVO_4) using light ions such as H and He+ and heavy ions such as C_2+. In general, these ions decrease the refractive index in the implanted area, producing a barrier that permits guiding in the region near the surface. Furthermore, information about nonlinear optical properties of channel waveguides containing metallic nanoparticles is presented. Composite materials containing metallic nanoparticles embedded in a dielectric matrix such as silica possess interesting properties due to surface plasmon resonance absorption features and the enhancement of the third-order nonlinear optical response. Therefore, nonlinear optical properties in composite waveguides can be used in all-optical switching devices.

Keywords: optical waveguides, multi-energy ion implantation

1. Introduction

Since the 1960s, waveguide optics has evolved into an emerging discipline and has a tremendous impact on our information society, due to the potential applications for signal processing

INTECH
open science | open minds

© 2017 The Author(s). Licensee InTech. This chapter is distributed under the terms of the Creative Commons Attribution License (http://creativecommons.org/licenses/by/3.0), which permits unrestricted use, distribution, and reproduction in any medium, provided the original work is properly cited. (cc) BY

and biochemical sensing [1]. The first proton implantation in fused silica to form waveguides was reported in 1968 [2]. Then, ion-implanted waveguide progress has been reviewed by Townsend et al. summarizing early results before 1993 [3]. In recent years, Chen et al. and Liu et.al. presented review papers [4–6] showing advances from 1994 to 2013. Ion implantation has been used to produce optical waveguides in several substrates, including more than 100 materials such as single crystals, glasses, polycrystalline ceramics and organic materials. Moreover, ion implantation allows accurate control of both dopant composition and penetration depth through the choice of the species and the energy of the ions. Waveguides can be fabricated at low temperature, which ensures steady chemical compositions and phases in the waveguide region. Waveguides can be fabricated by combining ion implantation with other techniques, such as ion exchange and metal ion diffusion [7, 8].

Single ion implantation has been used as a rule to produce optical waveguides, but there is a lack of control of optical confinement that limits its potential as optical waveguides [9]. However, optical waveguides can be improved by widening the barrier or directly generating the waveguide core through multiple energy ion implantations [10, 11]. There are a few works related to optical waveguides obtained by means of multiple ion implantations [12, 13]. In this chapter, multi-energy ion implantation is used to fabricate optical waveguides.

2. Theory

2.1. Optical waveguides

The simplest optical waveguide structure is the planar or slab waveguide. It is composed of a refractive index material surrounded by two lower refractive index materials, and it confines light in one direction. **Figure 1** shows the typical three-layer configuration of a planar waveguide, where n_1 is the refractive index of the core and n_2 and n_3 are the refractive indices of substrate and cladding, respectively ($n_1 > n_2 > n_3$). Considering a plane wave travelling in a slab waveguide structure, with a dielectric square shape core refractive index, an eigenvalue equation that describes light propagation can be obtained for transverse electric (TE) and transverse

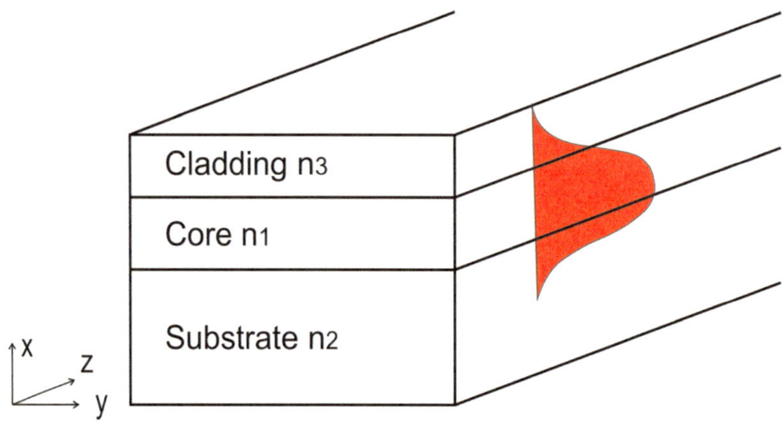

Figure 1. A three-layer configuration of a slab optical waveguide.

http://dx.doi.org/10.5772/67829

magnetic (TM) wave polarizations, Eqs. (1) and (2), for a three-layer dielectric waveguide structure [14]:

$$k_a a = tan^{-1}\frac{k_c}{k_a} + tan^{-1}\frac{k_s}{k_a} + m\pi \tag{1}$$

$$k_a a = \frac{1}{d}tan^{-1}\frac{k_c}{k_a} + \frac{1}{c}tan^{-1}\frac{k_s}{k_a} + m\pi \tag{2}$$

where a is the film physical thickness, $k_a = k\sqrt{n_a^2 - n_e^2}$, $k_s = k\sqrt{n_e^2 - n_s^2}$, and $k_c = k\sqrt{n_e^2 - n_c^2}$ are the transversal wavenumbers in the core, substrate, and cladding, respectively, and $k = 2\pi/\lambda$ is the wavenumber; $c = n_s^2/n_a^2$ and $d = n_c^2/n_a^2$ are parameters introduced in the TM eigenvalue equation. Furthermore, an effective index n_e can be obtained for every integer m number; this solution is known as the m_{th} mode and has its own light distribution at the waveguide structure. The distribution of the electric field amplitude is given by:

$$E_y(x) = \begin{cases} E_c e^{-\sigma_c x}, \ x \geq 0 \\ E_c\left[\cos(\kappa_a x) - \frac{\sigma_c}{\kappa_a}\sin(\kappa_a x)\right], \ -a \leq x \leq 0. \\ E_c\left[\cos(\kappa_a a) - \frac{\sigma_c}{\kappa_a}\sin(\kappa_a a)\right]e^{\sigma_s(x+a)}. \ x \leq -a \end{cases} \tag{3}$$

The light confinement, Γ, is defined as a ratio between the light power in the core and the total mode power and is calculated from [15]:

$$\Gamma = \frac{P_{film}}{P} = \frac{1 + \frac{k_s}{k_f^2 + k_s^2} + \frac{k_c}{k_f^2 + k_c^2}}{f + \frac{1}{k_s} + \frac{1}{k_c}}. \tag{4}$$

2.2. Ion implantation process

The ion implantation has proven to be a powerful technique to fabricate optical waveguides in a variety of materials. In the case of light ions that are implanted, the damage caused by the nuclear collisions during the implantation reduces the physical density of the substrate. The low-density buried region, with refractive lower than the substrate, acts as an optical barrier. The region between the surface and the optical barrier has a higher refractive index and can operate as a waveguide [12]. By contrast, heavy ions can increase physical density, polarizability or structure of the substrate, which results in an increase in the refractive index of the implanted region surrounded by regions with lower refractive index, making a typical optical waveguide [13].

Nuclear damage processes, which create partial lattice disorder in the material, also can introduce a positive or negative change in the refractive index, depending on the ionization and diffusion effects. The index profile depth can be controlled by the ions energy, and the positive or negative change in the refractive index will allow one to have a waveguide core or an optical barrier [3]. The changes in the refractive index n mostly depend on the defect content (ΔV), changes in the polarizability ($\Delta\alpha$) and structure factors (F). The Wei adaptation of the Lorentz-Lorenz equation gives [16]:

$$\frac{\Delta n}{n} = \frac{(n^2-1)(n^2+2)}{6n^2}\left[-\frac{\Delta V}{V} + \frac{\Delta \alpha}{\alpha} + F\right]. \quad (5)$$

When the dominant mechanism is a negative ΔV, an optical barrier will be produced beneath the substrate surface. On the other hand, a positive ΔV may lead to an increase in the refractive index acting as the core of the waveguide. Glass implanted with heavy ions such as Ag and Cu follows this behaviour. Furthermore, an annealing procedure can generate metal nanoparticles in the waveguide, which have potential as nonlinear optical devices [17].

3. Methodology and materials

The development of optical waveguides by a multi-implantation process shown in blocks in **Figure 2** requires feedback from optical design software, ion implantation simulations, optical characterization and optimization. Once specific requirements are defined for the design of optical waveguides, it is necessary to know basic information of the first approach of waveguide parameters such as Δn and core dimensions. Afterwards, it is necessary to try to replicate these features through the ion implantation process. The prediction of ion range and damage distribution is essential for the ion implantation technique and thus several computer simulation programs have been developed such as the Stopping and Range of Ions in Matter (SRIM), crystal- Transport of Ions in Matter (crystal-TRIM), Projected Range Algorithm (PRAL), MARLOWE, University of California MARLOWE (UC-MARLOWE), University of Valladolid MARLOWE (UVAMARLOWE), REED and so on [18–22]. Here, the ion trajectory was simulated by the Stopping and Range of Ions in Matter (SRIM) software [23], while the ions were implanted by the NEC Pelletron Accelerator Model 9SDH-2 at the Instituto de Física, UNAM. The experimental parameters used to fabricate SiO_2 planar waveguides by Ag+ multiple implantation process are shown in **Table 1**.

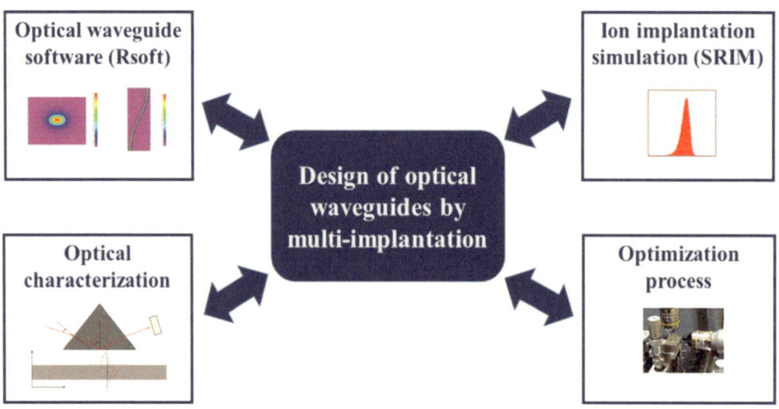

Figure 2. A diagram for the development of optical waveguides by a multi-implantation process.

http://dx.doi.org/10.5772/67829

Implant energy	Fluence [ions/cm²]			
	M1	M2	M3	M4
9 MeV	5×10^{14}	1×10^{15}	2.5×10^{15}	5×10^{15}
7 MeV	2.5121×10^{14}	5.025×10^{14}	1.256×10^{15}	2.5119×10^{15}
6 MeV	1.4748×10^{14}	2.9505×10^{14}	7.3757×10^{14}	1.475×10^{15}
5.2 MeV	2.0689×10^{14}	4.1375×10^{14}	1.034×10^{14}	2.069×10^{15}
4.3 MeV	1.9508×10^{14}	3.9×10^{14}	9.75×10^{14}	1.95×10^{15}

Table 1. Parameters of implantation used for the samples M1–M4 [13].

An automatic prism-coupling system, (Metricon Model 2010) operating at a wavelength of 633 nm, was used for obtaining TE/TM effective refractive indices of the waveguides propagation modes, and propagation losses were determined by a transmission technique; both techniques are described later.

Also, results will be reviewed for the waveguides fabricated in laser crystals such as YAG and YVO_4 doped with Nd or Yb. These results deal mainly with refractive index changes and laser emission properties.

4. Results and discussions

4.1. Initial parameters for optical waveguides

The single mode of behaviour of the waveguide in the Near Infrared (NIR) was considered as a major goal. Calculations by means of the Reflectivity Calculation Method (RCM) result in a step-index waveguide with a core width of ~3 μm and index change of Δn~0.008, whose intensity mode profiles are shown in **Figure 3**.

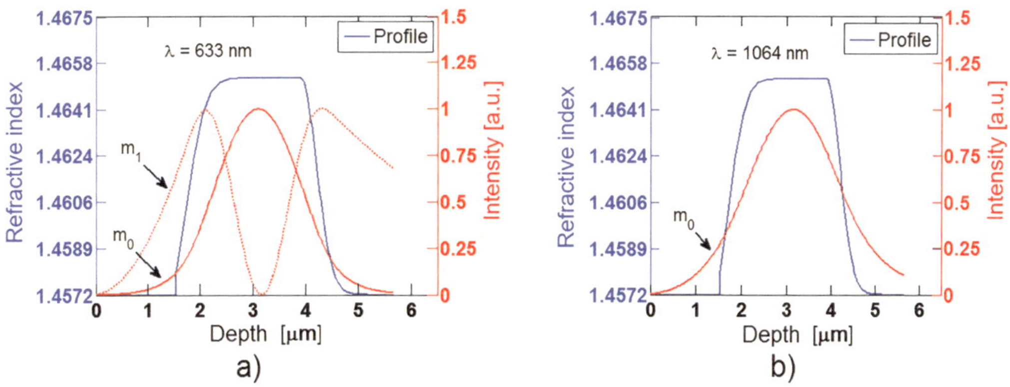

Figure 3. Intensity distribution of propagation modes confined in a step-index waveguide with refractive index change $\Delta n\sim0.008$ for: (a) $\lambda = 633\ nm$ and (b) $\lambda = 1064\ nm$.

4.2. Design of waveguides through multiple energy ion implantations

The methodology oriented to design optical waveguides in a dielectric matrix, by means of ion implantation, must consider parameters as: (a) type of ion, energy, fluence and (b) composition and structure of the substrate trying to obtain a specific waveguide. Moreover, changes in the properties of the substrate during the ion implantation, for example, structural, density or polarizability must be considered for a robust waveguide design.

4.2.1. Optical waveguides by heavy (Ag) ion implantation

First, the silver ion implantation profile was calculated using energies from 1 to 10 MeV and fluences from 10^{14} to 10^{15} ions/cm^2 in an SiO_2 substrate; see **Figure 4(a)** [17]. As an example, **Figure 4(b)** shows a step ion distribution obtained from multiple silver implantation energies and fluences: 9 MeV, 5x10^{14} ions/cm^2; 7 MeV, 2.5x10^{14} ions/cm^2; 6 MeV, 1.47x10^{14} ions/cm^2; 5.2 MeV, 2x10^{14} ions/cm^2 and 4.3 MeV, 1.95x10^{14} ions/cm^2 [13].

4.2.2. Optical waveguides by light (H, He+) ion implantation

Light ion implantation normally reduces the refractive index at the end of the ion trajectory and thus an optical waveguide is created between this low index region (so-called optical barrier) and the air next to the substrate's surface, see **Figure 5**. The refractive index is reduced mainly because of a volume expansion in the damage region, producing a lower material density. Depending on the material, the refractive index in the guiding region may increase or decrease slightly due to polarizability and structural factors, according to Eq. (5). In order to reduce light leakage through the optical barrier into the substrate (i.e., light tunnelling), generally, two or three energies close to each other are used, generating broader barriers than those obtained by a single implant [24–26]. For example, helium ion energies of 1.75, 1.6 and

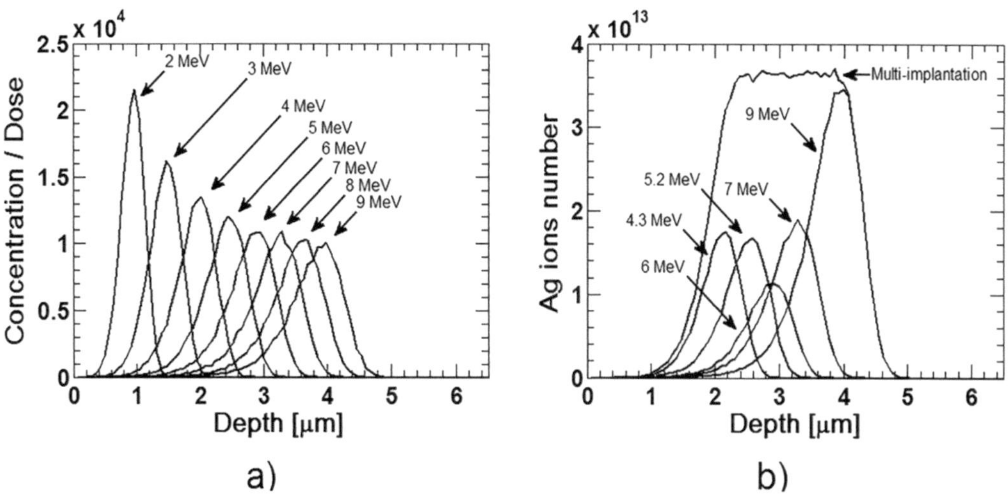

Figure 4. (a) Ag implantation profiles for a range of 2–9 MeV and (b) multiple implantations towards a generation of a step-index waveguide [13].

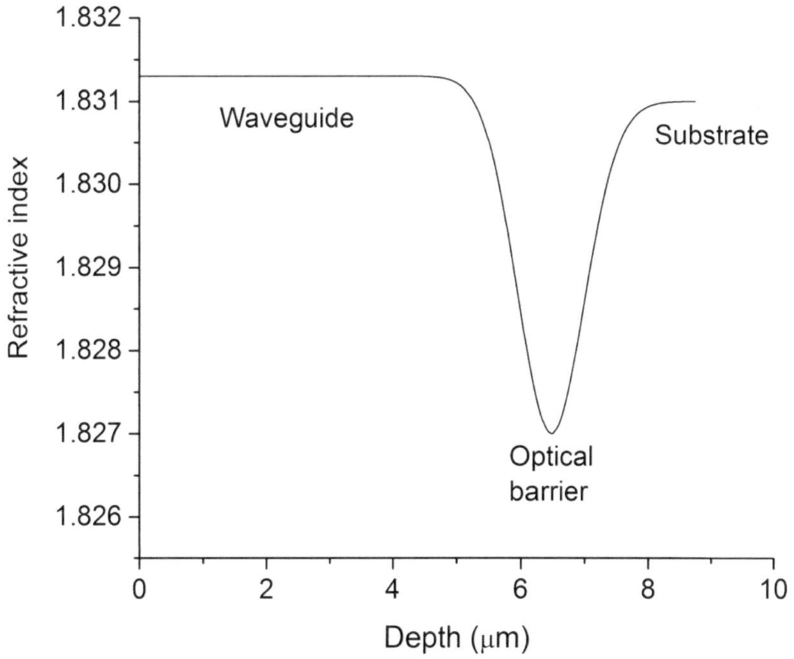

Figure 5. A refractive index profile of a Yb:YAG waveguide implanted with protons at an energy of 1 MeV and an incident angle of 46° [27].

1.5 MeV can be implanted, instead of only 1.5 MeV, at a dose of around 10^{16} ions/cm^2, reducing the tunnelling losses in the waveguide.

4.3. Passive optical waveguides

In this section, the results of optical and physical properties of ion-implanted waveguides are presented.

4.3.1. Refractive index change for an optical waveguide

One of the most important optical properties of waveguides is the refractive index change that is produced by the ion implantation process. From the complex mechanism of effects produced by ion implantation in optical materials, here, only the main contributors to the refractive index change are considered. Refractive index change is induced by the modification of the substrate density, polarizability or structure produced by radiation damage or stoichiometric changes [9–11]. An approach to analyse the refractive index change, Δn, of an ion-implanted waveguide is given by

$$\Delta n^{TE,TM} = \Delta n_R + \Delta n_V + \Delta n_\sigma^{TE,TM} \tag{6}$$

where Δn_R is mainly due to the difference in polarizability before and after ion implantation which can be calculated using a basic model that relates refractive index as a function of chemical composition [9]. Refractive index change's contribution from volume change of the matrix surface caused by the induced compaction process, Δn_V, can be estimated across the surface implanted through AFM measurements [28]. The index change Δn_σ depends on the polarization state of the light (TE or TM) is related to the optical stress produced by ion implantation which can be determined directly using an effective index of propagation modes for TE and TM polarizations.

An initial consideration to estimate the gradient index profile of the ion implantation distribution is to calculate the ion distribution generated by the multiple ion implantation process and the corresponding refractive index. First, we calculated Δn of each segment of 70 nm of the waveguide, shown in **Figure 6(a)**. Refractive index profiles obtained for M1-M4 are shown in **Figure 6(b)**. The values of Δn_{Rmax} calculated for samples *M1-M4* are ~0.00017, ~0.00034, ~0.00089 and ~0.0017, considering a silver ion concentration of Ag_{max} ~0.04%*wt.*, ~0.08%*wt.*, ~0.2%*wt.* and ~0.4%*wt.*, respectively.

4.3.2. *Refractive index profile of optical waveguides*

A way to determine the refractive index profile of an optical waveguide is by means of direct measurements of effective refractive indices of confined and radiated propagation modes using a prism-coupling technique. In the prism-coupling method, as shown in **Figure 7**, an incident light beam enters the prism at an angle φ. At the prism base, the light beam forms an angle θ to the normal. This angle, θ, determines the phase velocity in the z direction of the incident beam in the prism and in the gap between the prism and the waveguide. Efficient coupling of light into the waveguide occurs only when we choose the angle φ such that v_i is equal to the phase velocity, v_m, of one of the guided modes [29].

The effective refractive index, *Nm*, of the *m*th mode is related to θ_m, by

$$N_m = n_p sin\left[sin^{-1}\left(sin\frac{\theta_m}{n_p}\right) + A\right] \tag{7}$$

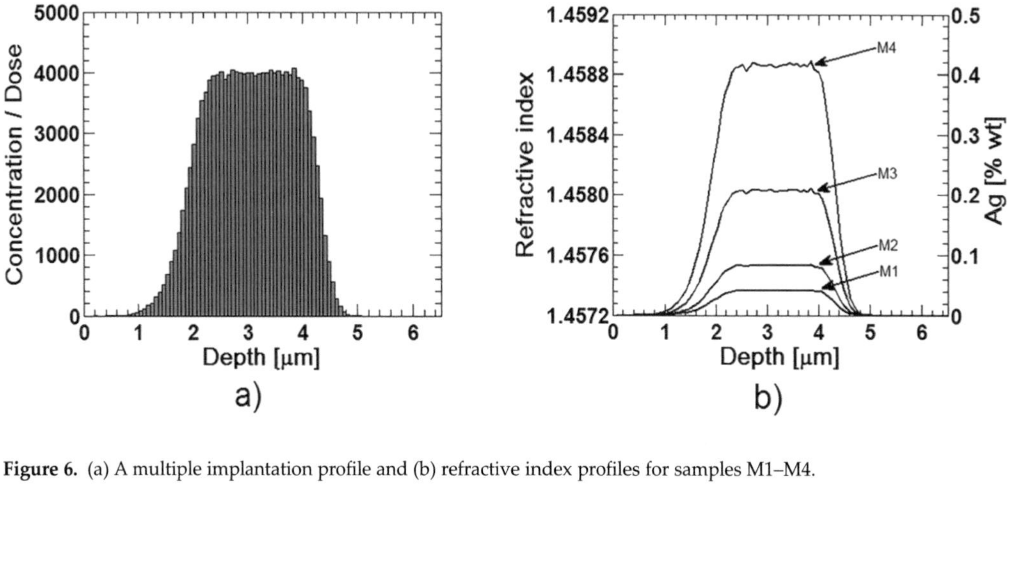

Figure 6. (a) A multiple implantation profile and (b) refractive index profiles for samples M1–M4.

http://dx.doi.org/10.5772/67829

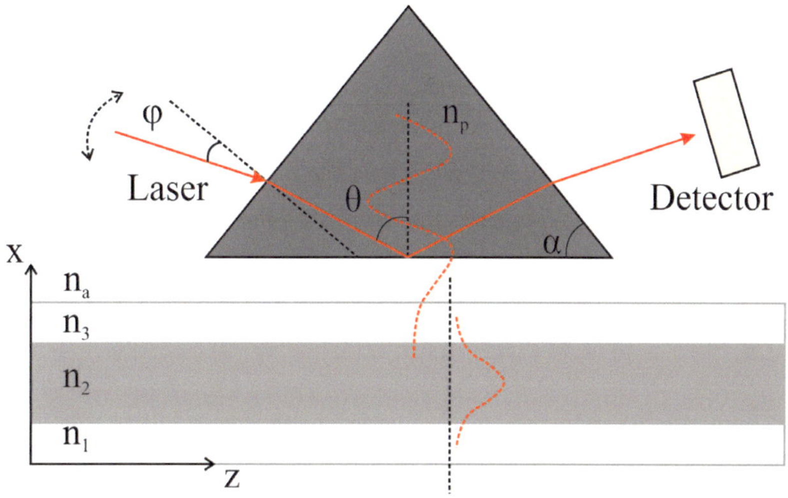

Figure 7. A principle of operation of the prism-coupling technique [29].

where A and n_p are the base angle and refractive index of the prism, respectively. The effective refractive indices of the guided modes in the SiO_2 waveguides, shown in **Table 2**, were obtained by the measurement of the coupling mode angles of the prism coupler. **Figure 8(a)** shows typical results of TE/TM effective refractive indices for the propagation modes of waveguides obtained through the prism-coupling technique. **Table 2** lists the values of TE and TM effective refractive indices for the propagation modes in the ion-implanted waveguides. Propagation modes m_0 and m_1 are guided modes and m_2, m_3 and m_4 are considered radiated modes because they do not fulfil the condition $n_3 < n_{eff} < n_2$ but are required for an adequate refractive index profile fitting [30]. Starting from a refractive index profile calculated from polarizability shown in **Figure 6(b)**, a refractive index profile for the waveguides was fitted by the RCM method for experimental and theoretical effective refractive indices of the waveguide propagation modes. Fitted refractive index profiles of waveguides, for TE/TM polarizations, are shown in **Figure 8(b)**, wherein values of refractive index from core, cladding and substrate are given, respectively.

Mode	**M1**		**M2**		**M4**	
	n_{effTE}	n_{effTM}	n_{effTE}	n_{effTM}	n_{effTE}	n_{effTM}
m_0	1.4640	1.4644	1.4642	1.4646	1.4640	1.4644
m_1	1.4604	1.4607	1.4607	1.4611	1.4593	1.4597
m_2	1.4542	1.4548	1.4548	1.4551	1.4518	1.4521
m_3	1.4428	1.4430	1.4431	1.4433	1.4384	1.4387
m_4	1.4275	1.4273	1.4287	1.4291	1.4192	1.4187

Table 2. Effective refractive indices of the propagation modes in the SiO_2 waveguides [13].

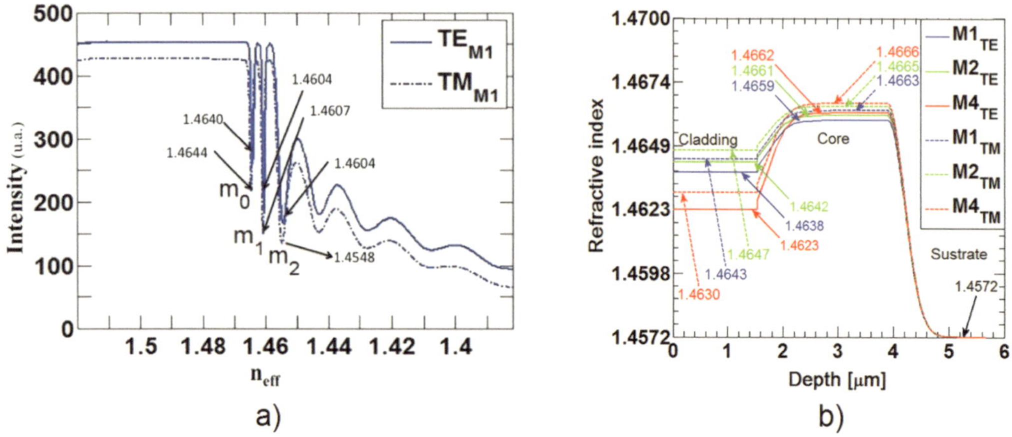

Figure 8. (a) An experimental TE/TM effective refractive index for the propagation modes of the waveguide M1 and (b) TE/TM refractive index profiles calculated for waveguides M1, M2 and M4 [13].

4.3.3. Waveguide propagation losses

Propagation losses were determined by the method of light transmission, as shown in **Figure 9**. Transmittance of the waveguides (t_w) was calculated considering losses along the different components involved in the whole system. Interface-waveguide transmittance ($t_{f-w} = 1 - \eta_R$), where η_R is the Fresnel reflection fibre guide, the mode-size mismatch between the fibre and the waveguide (η_0), interface transmittance (t_{out}) and microscope transmittance (t_m) are needed to be accounted in order to describe the overall throughput T, given by Eq. (8) [9].

$$T = t_{f-w}\eta_0 t_w t_{out} t_m \tag{8}$$

To determine the propagation losses (α_w), we use the extinction coefficient given by Eq. (9), where P_{in} and P_{out} are the input and output of optical power respectively and L is the waveguide length.

$$\alpha_w = -\ln\left(\frac{P_{in}}{P_{out}}\right)/L \tag{9}$$

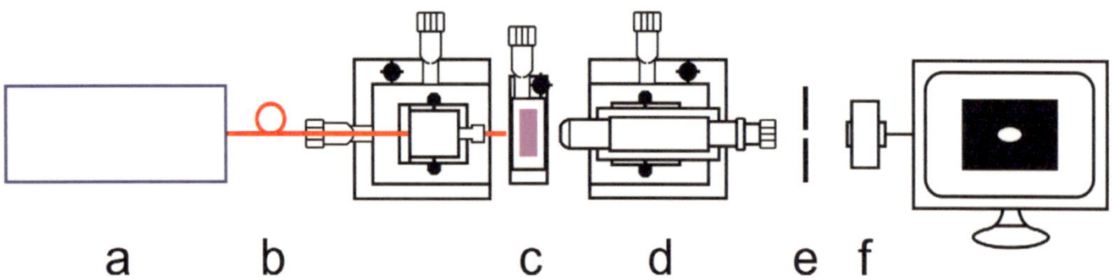

Figure 9. An experimental setup for a waveguide mode profile: (a) laser, (b) optical fibre, (c) waveguide, (d) microscope, (e) diaphragm and (f) CCD or detector.

To estimate losses and transmittance, power measurements were done by fibre-coupling into one waveguide facet, which introduces mainly three loss mechanisms: Fresnel reflections, size mismatch between the fibre mode and the waveguide mode and their misalignment [31].

The Fresnel reflection depends on the index change at both the waveguide and fibre end face. The reflected light (returned loss) can be written as:

$$\eta_R = \frac{\left(\frac{n_1-n_2}{n_1+n_2}\right)^2 + \left(\frac{n_3-n_2}{n_3+n_2}\right)^2 + 2\left(\frac{n_1-n_2}{n_1+n_2}\right)\left(\frac{n_3-n_2}{n_3+n_2}\right)\cos(4\pi n_2 g \frac{1}{\lambda})}{1 + \left(\frac{n_1-n_2}{n_1+n_2}\right)^2\left(\frac{n_3-n_2}{n_3+n_2}\right)^2 + 2\left(\frac{n_1-n_2}{n_1+n_2}\right)\left(\frac{n_3-n_2}{n_3+n_2}\right)\cos(4\pi n_2 g \frac{1}{\lambda})} g \tag{10}$$

where n_1 and n_3 are the effective refractive indices of the fibre and the guide mode, n_2 is the refractive index in the gap g between the fibre and the guide and λ is the wavelength. The mode-size mismatch induces loss because the transverse mode coupling cannot be complete. The efficiency is computed using the well-known overlap integral

$$\eta_0 = \frac{\iint |E_1 E_2|^2 dxdy}{\iint |E_1|^2 dxdy \iint |E_2|^2 dxdy}. \tag{11}$$

This integral expresses the coupling between the electric fields E_1 and E_2 of modes, which propagates in the fibre and in the waveguide, respectively. In the most general case, the mode profile can be approximated by a combination of half Gaussians (see **Figure 10**) and the overlap integral gives:

$$\eta_0 = \frac{\left(\sqrt{\omega_1}\left(\frac{\omega_1}{\omega_0} + \frac{\omega_0}{\omega_1}\right)^{-1/2} + \sqrt{\omega_2}\left(\frac{\omega_2}{\omega_0} + \frac{\omega_0}{\omega_2}\right)^{-1/2}\right)^2}{\frac{\omega_1+\omega_2}{2}\left(\frac{\omega_3}{\omega_0} + \frac{\omega_0}{\omega_3}\right)} \tag{12}$$

where ω_0 is the waist of the fibre mode and ω_1, ω_2, and ω_3 are the half-waists of the waveguide mode. The propagation loss values (α_w) measured for the waveguides M_1, M_2 and M_4 are around 0.7, 3.9 and 19.5 dB/cm, respectively.

Figure 11 shows the intensity distribution of the propagation modes supported by the silver-implanted waveguides in SiO_2. These images of the spot emerging from the waveguide output, excited by fibre-waveguide coupling, were obtained by a Charge-coupled device (CCD) camera coupled to a travelling microscope. Here, it is possible to appreciate the excitation of propagation modes m_0 and m_1 for each of the waveguides, which is in accordance with results from the prism-coupling technique. The waveguide optical confinement factor measured for propagation modes m_0 and m_1 was 90% and 70% for M1, 80% and 75% for M2 and 90% and 85% for M4, respectively.

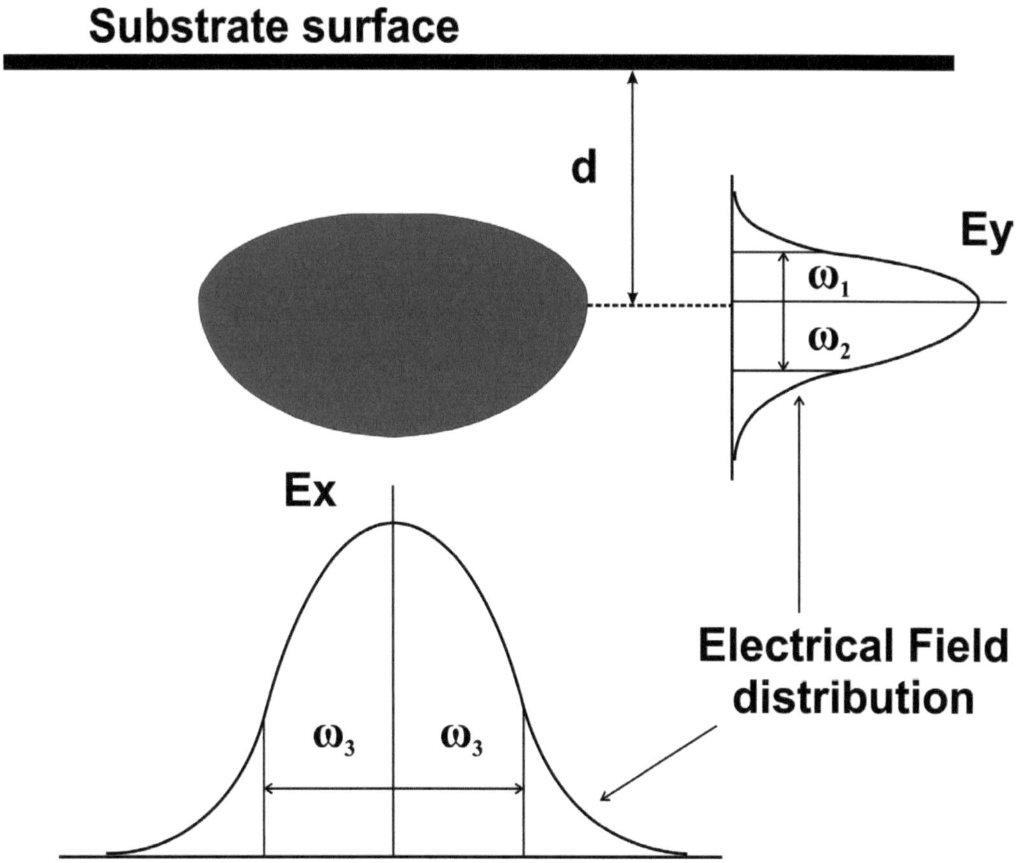

Figure 10. A typical waveguide mode profile [31].

4.4. Active optical waveguides

4.4.1. Laser emission

For more than 50 years, a large variety of lasers have been developed and optimized for different performance characteristics such as output power, efficiency, emission bands, pulse energy and pulse width. Research in this area is sustained by the impulse of a great diversity of scientific and technological applications in our modern society.

The main properties of laser emission are (a) coherence, which means that the photons move synchronously along the beam, (b) directionality, that is its ability to be focused into a small spot and (c) monochromaticity, which means that the light beam consists of essentially one wavelength, and this property originates from the basic principle of stimulated emission that involves well-defined atomic energy levels. Among different types of lasers, solid-state lasers are based on crystals or glasses doped with rare earth or transition metal ions. The principal source of energy to obtain efficient laser emission in these materials comes from laser diodes, giving solid-state lasers some advantages such as compact configuration, prolonged

http://dx.doi.org/10.5772/67829

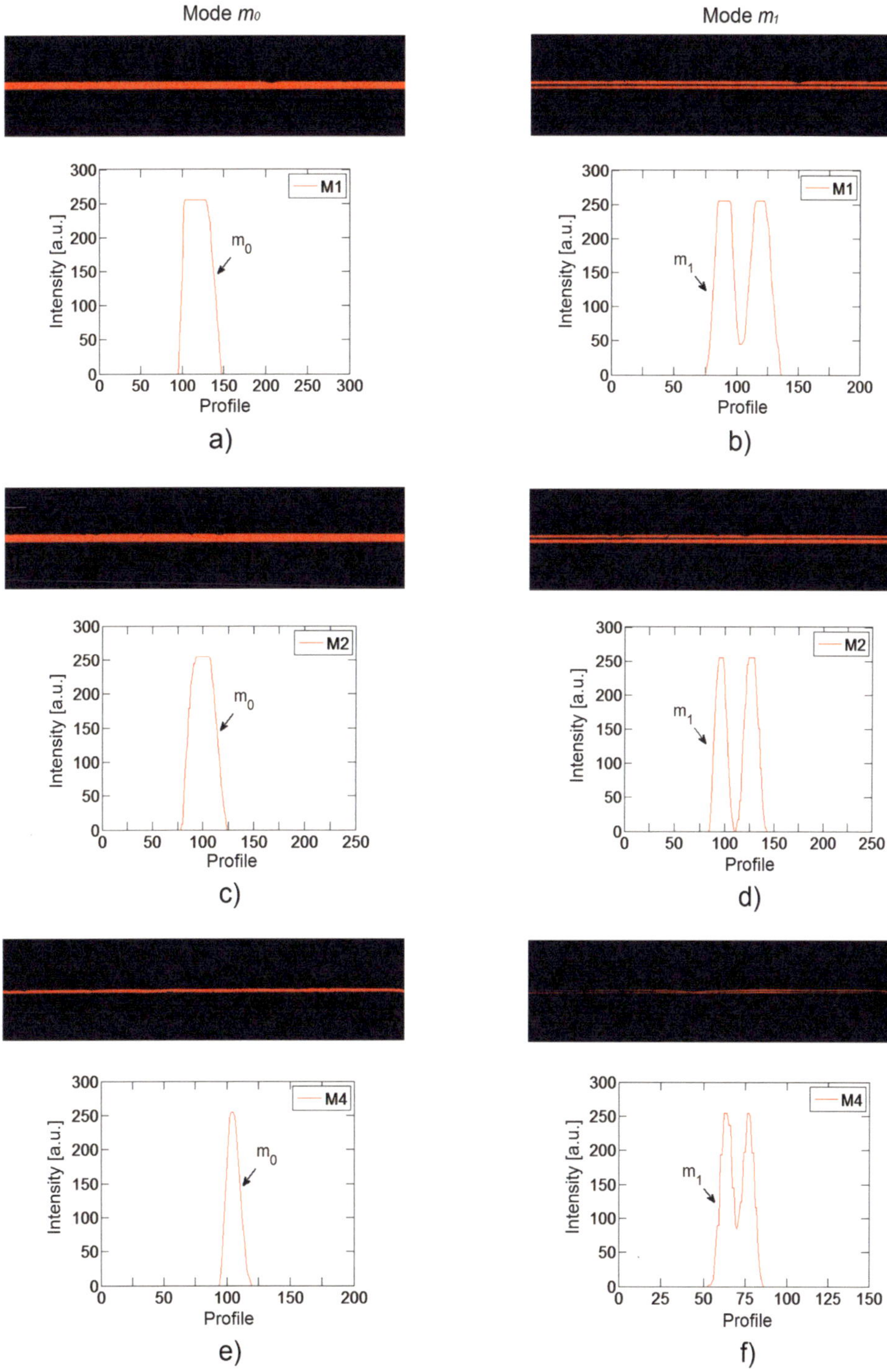

Figure 11. Intensity distributions of waveguide output light and intensity distribution profile for waveguides: a) fundamental mode in M1, b) first mode in M1, c) fundamental mode in M2, d) first mode in M2, e) fundamental mode in M4, and f) first mode in M4.

operational lifetime and generally very good beam quality. In particular, waveguide lasers offer high gain and low power thresholds in a small cross-section.

The principal laser characteristics are slope efficiency (ϕ) and threshold pump power (P_{th}); these are obtained from the plot of the laser output power as a function of the absorbed pump power (i.e., slope efficiency curve). The slope efficiency is the slope of the curve when it is approximated to a straight line, and the threshold pump power is the value of the pump power at which the curve crosses the pump power axis. For a four-level system like Nd:YAG or Nd:YVO_4, these parameters can be estimated using the following equations [32]:

$$\phi = \eta \frac{1 - R_2}{\delta} \frac{\lambda_p}{\lambda_s} \tag{13}$$

$$P_{th} = \frac{hc}{\lambda_p} \frac{1}{\eta \sigma_e \tau} \frac{\delta}{2} A_{eff} \tag{14}$$

where η is the pump quantum efficiency (number of ions excited to the upper-laser level per absorbed photon), $\delta = 2\alpha l - \ln(R_1 R_2)$ gives the total cavity losses (where α is the waveguide propagation loss and l is the cavity length), R_1 and R_2 are the reflectivities of the input and output mirrors, λ_p and λ_s are the pump and signal wavelengths, respectively, h is the Planck's constant, c is the speed of light in the vacuum, σ_e is the stimulated emission cross-section, τ is the fluorescence lifetime and A_{eff} is the effective pump area.

4.4.2. *Optical waveguides in active crystals*

Waveguides in laser and nonlinear crystals have been developed for several decades achieving a mature stage. In particular, rare earth-doped waveguides combine compactness and the possibility of generating laser light, by combining the absorption and emission properties of the active ion with the confinement capacity of the optical microstructure.

Numerous laser waveguides have been fabricated by ion implantation using the optical barrier approach (i.e., reducing the refractive index at some distance beneath the surface) [3–5]. In this case, multi-energy implantation has been used to broaden the barrier and thus reduce the light tunnelling into the substrate. Our group has reported laser waveguides in Nd:YAG, Yb:YAG and Nd:YVO_4 by light and heavy ion implantation [12, 27]. YAG waveguides exhibit a slight increase in the refractive index in the guiding region, besides the optical barrier, due to polarizability effects and optical stress. As an example of laser emission in these structures, channel waveguides fabricated by proton implantation in Nd:YAG showed a laser pump threshold of 6.9 mW and a slope efficiency of 16% [33], see **Figure 12**. Other groups have reported better laser performances (e.g., threshold pump power as low as 1.6 mW with a slope efficiency of 29%), indicating the improvement in the production of waveguide lasers [34].

Nd:YVO_4 waveguides have been fabricated by proton, helium, carbon and silicon implantation among different ions [26, 35, 36]. Implanting light ions generates the typical optical barrier waveguide for both the ordinary and the extraordinary index. We used double or triple implants to reduce the tunnelling losses. For example, a comparison was made between triple

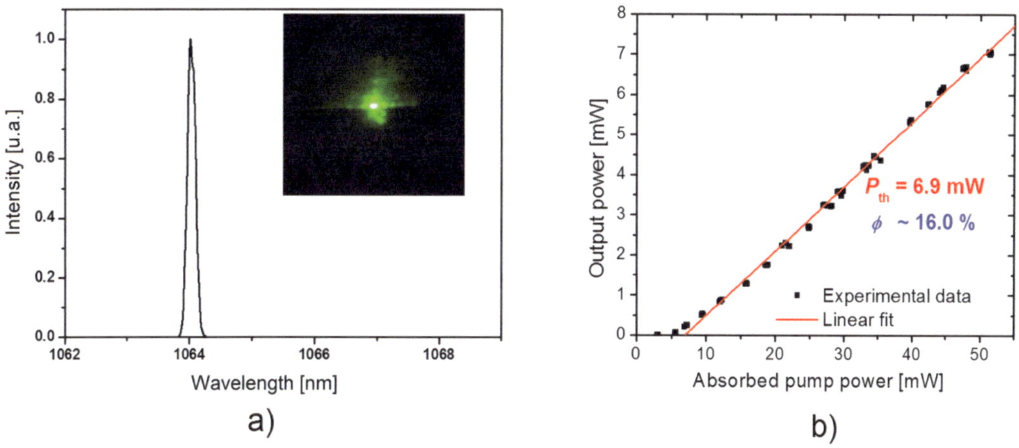

Figure 12. (a) Laser emission spectrum and (b) slope efficiency curve from a proton-implanted channel waveguide in Nd: YAG. The inset shows a photograph of the laser output at 1064 nm [33].

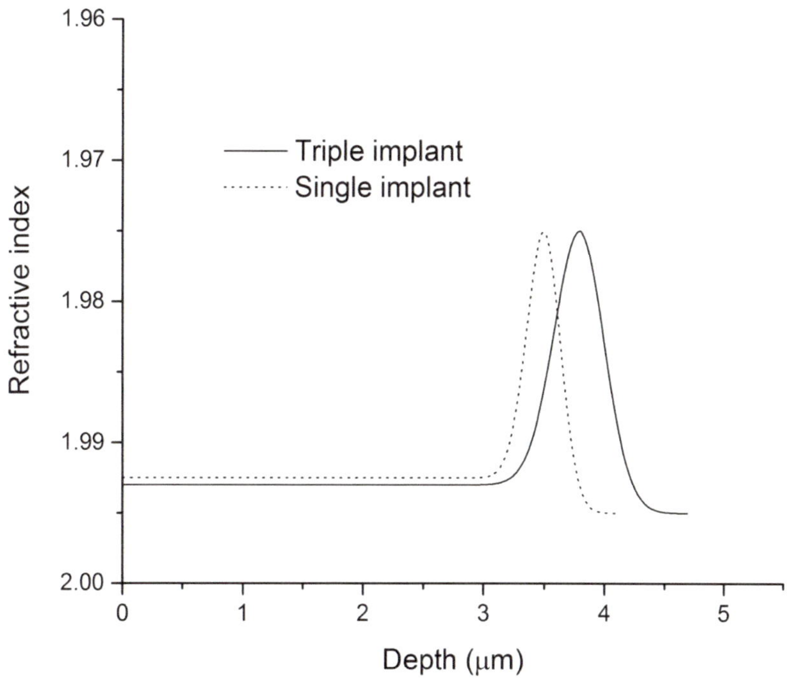

Figure 13. Refractive index profiles of Nd:YVO_4 waveguides implanted with a single implant and a triple helium implant [26].

helium-implanted and a single helium-implanted waveguide, see **Figure 13**; the light confinement was better in the multi-implanted waveguide due to the reduced tunnelling through the barrier. We also fabricated stacked waveguides in this crystal; in this case, we made two proton implants at 0.8 and 0.75 MeV to generate a deep broad optical barrier and a shallow implant at

0.4 MeV, thus producing a superficial and a buried waveguide. The buried waveguide had a better light confinement because the index contrast at the interface is smaller than in the shallow waveguide [26].

There are examples of materials that show a refractive index increase at certain light direction, thanks to their birefringence (e.g., YVO_4 and $LiNbO_3$). This property has been taken advantage in $LiNbO_3$ in order to produce a step-index profile for the extraordinary index by implanting carbon ions at 10 different energies [10]. However, it is possible to produce a refractive index increase with a single implant; this is realized by implanting heavy ions such as carbon and silicon at doses of around 10^{14} to 10^{15} ions/cm^2 [35, 36]. In this case, the refractive index profile exhibits an enhanced index well (i.e., the waveguide) and a region of reduced index at the end of the ion trajectory (i.e., the optical barrier). For a highly birefringent material, the electronic energy deposition causes lattice relaxation in the guiding region, increasing the lower index and reducing the higher one. We fabricated such waveguides (channels) in a Nd:YVO_4 crystal and demonstrated laser emission for the first time to our knowledge in a waveguide configuration. The threshold pump power was approximately 58 Mw, and the slope efficiency was around 24% [37].

4.5. Metallic nanoparticle-based optical waveguide

As it was previously described, ion implantation is a useful technique for producing waveguiding devices, either by using light ions to create a low index 'barrier' region or with metallic ions to increase the index in the implanted region and hence get guiding in this implanted region. On the other hand, it is well known that metal ions contained within a dielectric can be made to coalesce in order to form nanoparticles by thermal treatment in a reducing atmosphere [38]. The density and size of the nanoparticles can be controlled very well by controlling the implantation parameters such as dose and accelerating energy and by the thermal treatment parameters: temperature, duration and atmosphere composition. These nanocomposites containing metallic nanoparticles are very interesting from the point of view of their nonlinear optical properties, which are enhanced by the presence of localized surface plasmon resonance [39, 40].

The usual geometry for these samples is in the form of a thin layer, with thicknesses of the order of 1 μm, which are large enough to produce appreciable effects with low energy femtosecond pulses. It is therefore interesting to produce waveguides based on these metallic nanocomposite materials. The irradiance enhancement produced by the confinement of the electromagnetic field in a channel waveguide, and the considerably longer interaction length (from the mm to cm range), allows the observation of nonlinear effects with considerably smaller input pulse energies. Waveguide-based nonlinear devices are of particular interest for the implementation of ultrafast optical information processing applications [41] that exploit the nonlinear response to perform different functions. The fact that the region containing the metallic nanoparticles has a higher refractive index than the surrounding undoped silica substrate means that the implanted region can constitute the actual waveguide and that the nonlinear properties of the nanocomposite can be exploited for these applications, making them very good candidates for the implementation of such devices.

We have followed different approaches for the fabrication of channel waveguides containing metallic nanoparticles. In the first one, we have used a masked ion implantation to selectively implant Ag ions in narrow strips of the host, silica, followed by a thermal treatment to nucleate Ag nanoparticles [42]. As before, multiple implantations at different energies were employed to produce a cross-section that is large enough for efficient coupling by a single-mode fibre butt coupling. The energies employed in the successive implantations were 1.5, 2.0, 2.5, 3.0, 4.0 and 4.5 MeV. Because the widths and heights of the distribution vary with ion energy for a given dose, the doses at each energy were adjusted as to give the designed uniform profile. After the implantation process, the samples were thermally annealed at 600°C in a 50% N_2 + 50% H_2 reducing atmosphere for 1 hour, in order to obtain the highest amount of near-spherical nanoparticles by nucleation of the implanted ions. The end result is a sample containing channel waveguides with Ag nanoparticles, in a 2μm thick layer at a 0.52μm depth inside the silica matrix and having three different widths: 10,15 and 20 μm. **Figure 14** shows a white light microscope image of the waveguide and a front view of guided white light.

We characterized the linear optical properties of the waveguides produced using the techniques illustrated in **Figure 15(a)**. First, by looking at the light scattered perpendicular to the propagation direction, we can determine the propagation losses, and then by imaging the output at the guide end face, we can visualize the propagation modes of the guides. Propagation losses as low as 0.43 cm^{-1} were obtained for the 20 μm-wide guides using 633 nm light [42], and **Figure 15(b)** and **(c)** shows the measured modal field distribution, from where we can see that the guide supports three modes.

We have explored another approach for producing channel waveguides, which consists of direct writing of the waveguides by selective destruction of the nanoparticles using tightly

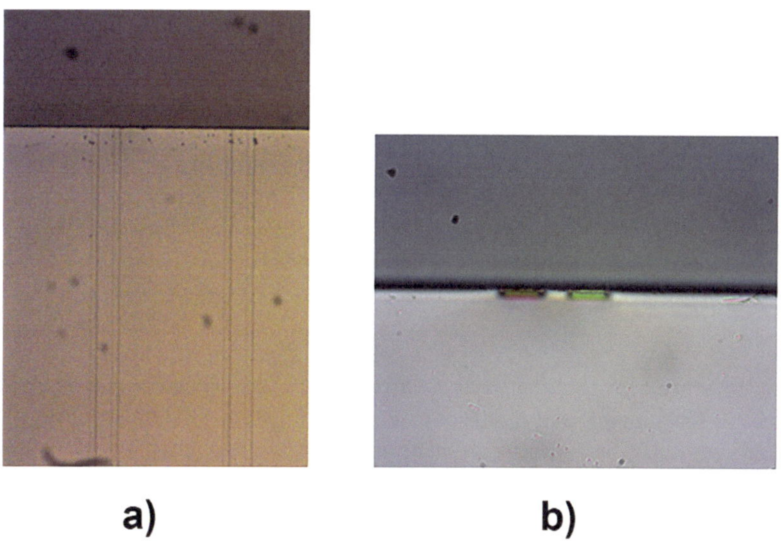

Figure 14. (a) A white light microscope image of an Ag nanoparticle waveguide (upper view) and (b) a front view of coupled white light output. The colouration is probably due to the plasmonic absorption of the nanoparticles.

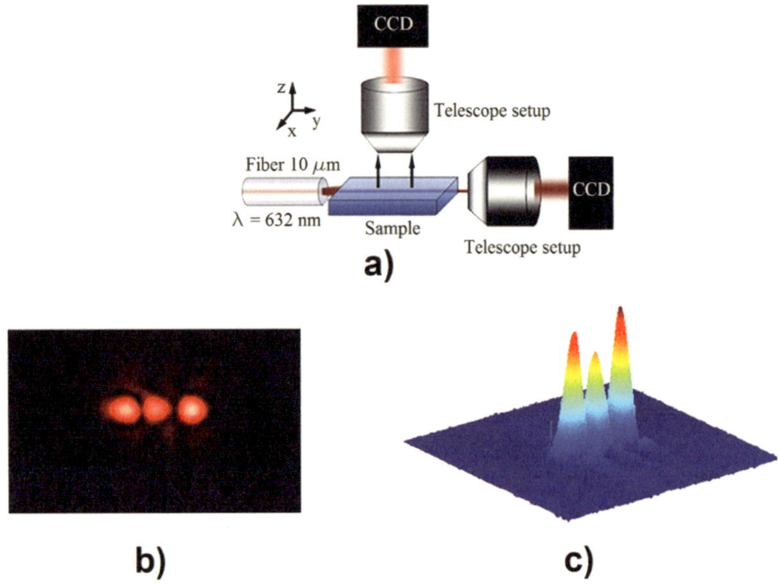

Figure 15. (a) An experimental set up to measure propagation losses and mode field distribution, (b) a 2D modal distribution measured for the 10 mm-wide guides and (c) an intensity plot for the measured mode.

focused *fs* pulses [43]. The advantage of this method is that by controlling the position of the laser pulses used, we can produce waveguides with different geometries and even more complex devices, such as Mach-Zehnder interferometers or directional couplers.

Based on our previous knowledge of the second harmonic generation properties of these materials [44], we employed a scanning second harmonic generation microscope based on an *fs* Titanium:sapphire laser oscillator, operated at high irradiances to produce the nanoparticle damage and at lower irradiances to image the damage achieved [43]. In this case, we used a sample that had elongated Ag nanoparticles that were aligned in a preferential direction, which was chosen to be at 45°C with respect to the sample surface [38]. This sample was chosen because of the relatively strong second harmonic signal observed [44] and the facility to induce damage. **Figure 16** shows a second harmonic image after we made two linear scans of the laser pulses at high intensity. It can be clearly seen that in the scanned regions, the second harmonic signal disappears almost completely, indicating the disappearance of the nanoparticles.

We have not observed waveguiding in these structures, but we expect to achieve it soon. This will allow us to produce different devices, such as beam splitters (Y junctions), Mach-Zehnder interferometers or directional couplers, which, together with the relatively large nonlinearities observed, can be used to implement all-optical switching devices, but of course, more work is needed in this direction.

Optical waveguides have been obtained through multiple energy ion implantations as a way to design optical waveguides with better optical confinement and lower propagation losses. The

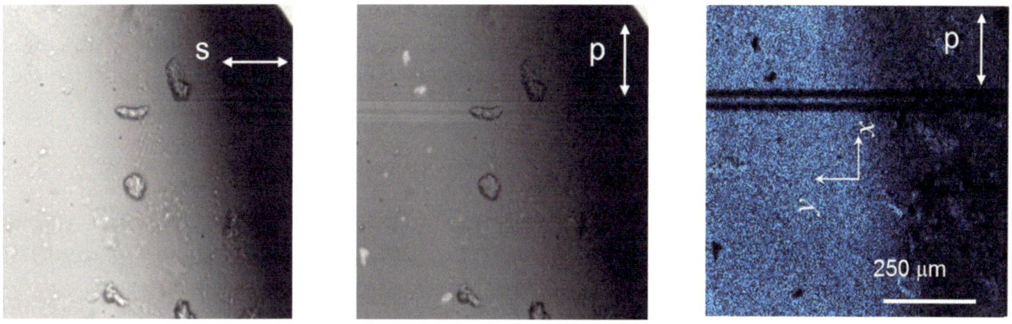

Figure 16. (a) Nanoparticle removal evidenced by the second harmonic generation and (b) white light microscopy images obtained with perpendicular polarizations, and (c) second harmonic image. Very high contrast is observed in the latter, indicating nanoparticle destruction.

protocol of design presented can be used with heavy ion implantation to enhance waveguide properties as core size and Δn; moreover, if the limit of metal solubility is reached, metallic nanoparticles can precipitate by means of annealing, bringing about a potential nonlinear optical waveguide. On the other hand, if light ions are used, it is possible to enhance the optical barrier or to design buried waveguides, that is, core waveguides formed between two barriers. This is possible due to the flexibility of the implantation process to change parameters as energy of implantation and fluence that must permit the development of active and passive integrated optical circuits.

5. Conclusions

Multiple ion implantation processes provide a potential way to produce optical waveguides using heavy and light ions, which improve capabilities to enhance optical confinement and reduce propagation losses—both paramount properties for optical waveguides. The protocol of design presented can be used with heavy ion implantation to enhance the core waveguide properties such as core size and Δn; moreover, if the limit of solubility of metal is reached, metallic nanoparticles can be produced by means of annealing, triggering a potential nonlinear optical waveguide. Ion implantation induces changes in polarizability, damage, compaction and stress that contribute directly to the refractive index profile and needs to be considered for advanced applications as active and passive integrated optical circuits.

Acknowledgements

The authors are grateful to CONACYT Grants No. 222485, No. J42695-F, No. 154021-F and DGAPA-UNAM IN-108217. The authors thank K. López and J.F. Jaimes for the IFUNAM Pelletron accelerator operation and Jessica Lilian Angel-Valenzuela for optical measurements support.

Author details

Heriberto Márquez Becerra[1]*, Gloria V. Vázquez[2], Eder G. Lizárraga-Medina[1], Raúl Rangel-Rojo[1] and David Salazar[1] and Alicia Oliver[3]

*Address all correspondence to: hmarquez@cicese.mx

1 Department of Optics, Center for Scientific Research and Higher Education at Ensenada, CICESE, Ensenada Baja California, México

2 Research Center in Optics, CIO, León Guanajuato, México

3 Institute of Physics, National Autonomous University of Mexico, UNAM, Mexico City, México

References

[1] Wang X, Ying C, Cao Z. Progress in planar optical (1st ed.). Berlin and Heidelberg: Springer-Verlag; 2016. 241 p. doi:10.1007/978-3-662-48984-0

[2] Schineller ER, Flam RP, Wilmot DW. Optical waveguides formed by proton irradiation of fused silica. J. Opt. Soc. Am. 1968;**58**(9):1171–1173

[3] Townsend PD, Chandler PJ, Zhang L. Optical effects of ion implantation (1st ed.). Cambridge: Cambridge University Press; 1994. 296 p

[4] Chen F, Wang X, Wang K. Development of ion implanted optical waveguides in optical materials: a review. Opt. Mater. 2007;**29(**11):1523–1542

[5] Chen F. Micro-and submicrometric waveguiding structures in optical crystals produced by ion beams for photonic applications. Laser Photon. Rev. 2012;**6**(5):622–640

[6] Liu CX, Peng B, Wei W, Li WN, Guo HT, Cheng S. Ion-implanted glass waveguides: a review. Phys. Int. 2013;**4**(1):1–12

[7] Tong XC. Advanced materials for integrated optical waveguides (1st ed.). Cham: Springer International Publishing; 2014. 552 p. doi: 10.1007/978-3-319-01550-7

[8] Xiang B, Ren X, Ruan S, Wang L, Yan P, Han H, Wang M, Yin J. Visible to near-infrared supercontinuum generation in yttrium orthosilicate bulk crystal and ion implanted planar waveguide. Sci. Rep. 2016;**6**:31612. doi:10.1038/srep31612

[9] Márquez H, Salazar D, Rangel-Rojo R, Angel-Valenzuela JL, Vázquez GV, Flores-Romero E, Rodríguez-Fernández L, Oliver A. Synthesis of optical waveguides by silver ion implantation. Opt. Mater. 2013;**35**(5):927–934

[10] Montanari GB, De Nicola P, Sugliani S, Menin A, Parini A, Nubile A, Bellanca G, Chiarini M, Bianconi M, Bentini GG. Step-index optical waveguide produced by multistep ion implantation in $LiNbO_3$. Opt. Express. 2012;**20**(4):4444–4453

[11] Ramabadran UB, Jackson HE, Boyd JT. Optical waveguides fabricated by ion implantation of Si^+, and N^+ in SiO_2: a Raman investigation. Appl. Opt. 1993;**32**(3):313–317

[12] Vazquez GV, Marquez H, Flores-Romero E, Sanchez-Morales ME. Optical waveguides fabricated by ion implantation in laser crystals. In: Espinoza-Luna R, Bernabeu E, Aboites V, editors. Recent research in photonics (1st ed.). Kerala: Research Signpost; 2009, pp. 153–181

[13] Ramírez-Espinoza C, Salazar D, Rangel-Rojo R, Angel-Valenzuela JL, Vázquez GV, Flores-Romero E, Rodríguez-Fernández L, Oliver A, Domínguez DA, Márquez H. Design of step-index optical waveguides by ion implantation. J. Lightwave Technol. 2015;**33** (14):3052–3059

[14] Lizarraga-Medina EG, Salazar D, Vázquez GV, Salas-Montiel R, Nedev N, Márquez H. Study of SiOx ($1 < \times < 2$) thin-film optical waveguides. J. Lightwave Technol. 2016;**34** (21):4926–4932

[15] Chen CL. Step-index thin film waveguides. In: Foundations for guided-wave optics (1st ed.). Hoboken: Wiley; 2007, pp. 25–49

[16] Wei DT, Lee WW, Blooom LR. Large refractive index change induced by ion implantation in lithium niobate. Appl. Phys. Lett. 1974;**25**(6):329–331

[17] Márquez H, Salazar D, Rangel-Rojo R, Angel JL, Vázquez GV, Flores-Romero E, Rodríguez-Fernández L, Oliver A. Waveguides by multiple implantations of Ag ion on SiO_2 substrates. In: Rodríguez-Vera R, Díaz-Uribe R, editors. 22nd Congress of the International Commission for Optics: Light for the Development of the World; October; Puebla. SPIE; 2011. doi:10.1117/12.901975

[18] Hernández-Mangas JM. Simulación de la implantación iónica en semiconductores [thesis]. Valladolid: Universidad de Valladolid; 2000, 192 p

[19] Biersack J.P., Haggmark L.G. A Monte Carlo computer program for the transport of energetic ions in amorphous targets. Nucl. Instr. and Meth. in Phys. Res. B. 1980;**174**(1–2): 257–269

[20] Ziegles J.F., Biersack J.P. The stopping and range of ions in matter. In: Bromley D.A., editor. Treatise on Heavy-Ion Science. New York: Springer US; 1985. p. 93–129. DOI: 10.1007/978-1-4615-8103-1_3

[21] Webb RP. Computer codes and simulation background to ion implantation distribution and sputtering programs. In: Briggs D, Seah MP, editors. Practical surface analysis: ion and neutral spectroscopy (2nd ed.). Hoboken: John Wiley & Son Ltd; 1992, pp. 657–704

[22] Posselt M. Crystal-trim and its application to investigations on channeling effects during ion implantation. Rad. Effects. 1994;**130**(1):87–119

[23] Ziegler J.F., Computer code SRIM. Available: http://www.srim.org in March 2017.

[24] Zhao JH, Du J, Wang FX, Qin XF, Fu G. Planar Nd:YGG waveguide fabrication by multiple energy He ion implantation. Nucl. Instrum. Meth. 2014;**326**:95–98

[25] Liu T, Liu P, Zhang L, Zhou YF, Yu XF, Wang XL. Planar waveguide in beta barium borate formed by proton implantation and optical properties in visible and near-infrared band. Opt. Mater. 2013;**35**(12):2068–2071

[26] Sánchez-Morales ME, Vázquez GV, Moretti P, Márquez H. Optical waveguides in Nd: YVO_4 crystals by multi-implants with protons and helium ions. Opt. Mater. 2007;**29** (7):840–844

[27] Vázquez GV, Ramírez D, Márquez H, Flores-Romero E, Rickards J, Trejo-Luna R. Waveguiding properties in Yb:YAG crystals implanted with protons and carbon ions. Appl. Opt. 2012;**51**(22):5573–5578

[28] Oven R. Polarizability, volume expansion, and stress contributions to the refractive index change of Cu^+-Na^+ ion exchanged waveguides in glass. Appl. Opt. 2011;**50**(26):5073–5079

[29] Najafi SI. Introduction to glass integrated optics (1st ed.). Norwood: Artech House Publishers; 1992, 180 p

[30] Chandler PJ, Lama FL. A new approach to the determination of planar waveguide profiles by means of a non-stationary mode index calculation. Opt. Acta. Int. J. Opt. 1986;**33** (2):127–143

[31] Bourhis J.F. Fiber-to-waveguide connection. In: S.I. Najafi, editor. Glass Integrated Optics and Optical Fiber Devices, Critical Reviews of Optical Science and Technology, SPIE Optical Engineering Press, CR53. San Diego, Ca., 1994, p. 335–366

[32] Clarkson WA, Hanna DC. Effects of transverse-mode profile on slope efficiency and relaxation oscillations in a longitudinally-pumped laser. J. Modern Opt. 1989;**36**(4):483–498

[33] Flores-Romero E, Vázquez, GV, Márquez H, Rangel-Rojo R, Rickards J, Trejo-Luna R. Laser emission in proton-implanted Nd:YAG channel waveguides. Opt Express. 2007;**15** (26):17874–17880

[34] Field SJ, Hanna DC, Large AC, Shepherd DP, Tropper AC, Chandler PJ, Townsend PD, Zhang L. Low threshold ion-implanted Nd:YAG channel waveguide laser. Electron. Lett. 1991;**27**(25):2375–2376

[35] Vázquez GV, Sánchez-Morales ME, Márquez H, Rickards J, Trejo-Luna R. Analysis of ion implanted waveguides formed on Nd:YVO_4 crystals. Opt. Commun. 2004;**240**(4–6): 351– 355

[36] Fu G, Wang KM, Wang XL, Lu F, Lu QM, Shen DY, Ma HJ, Nie R. Formation of planar optical waveguide by multi energy Si ion implantation into Nd:YVO_4 crystal. Surf. Coat. Technol. 2007;**201**(9–11):5427–5430

[37] Sánchez-Morales ME, Vázquez GV, Mejía EB, Márquez H, Rickards J, Trejo-Luna R. Laser emission in Nd:YVO_4 channel waveguides at 1064 nm. Appl. Phys. B. 2009;**94**(2):215–219

[38] Oliver A, Reyes-Esqueda JA, Cheang-Wong JC, Román-Velázquez CE, Crespo-Sosa A, Rodríguez-Fernández L, Seman JA, Noguez C. Controlled anisotropic deformation of Ag nanoparticles by Si ion irradiation. Phys. Rev. B. 2006;**74**(26):245425

[39] Hamanaka Y, Hayashi N, Nakamura A, Omi S. Dispersion of third-order nonlinear optical susceptibility of silver nanocrystal-glass composites. J. Lumin. 2000;**87**:859–861

[40] Rangel-Rojo R, McCarthy J, Bookey HT, Kar AK, Rodriguez-Fernandez L, Cheang-Wong JC, Crespo-Sosa A, Lopez-Suarez A, Oliver A, Rodriguez-Iglesias V, Silva-Pereyra H. Anisotropy in the nonlinear absorption of elongated silver nanoparticles in silica, probed by femtosecond pulses. Opt. Commun. 2009;**282**(9):1909–1912

[41] Paquot Y, Schröder J, Van-Erps J, Vo TD, Pelusi MD, Madden S, Luther-Davies B, Eggleton BJ. Single parameter optimization for simultaneous automatic compensation of multiple orders of dispersion for a 1.28 Tbaud signal. Opt. Express. 2011;**19**(25):25512–25520

[42] Can-Uc B, Rangel-Rojo R, Marquez H, Rodríguez-Fernández L, Oliver A. Nanoparticle containing channel waveguides produced by a multi-energy masked ion-implantation process. Opt. Express. 2015;**23**(3):3176–3185

[43] Licea-Rodríguez J, Rocha-Mendoza I, Rangel-Rojo R, Rodríguez-Fernández L, Oliver A. Femtosecond laser writing over silver nanoparticles system embedded in silica using nonlinear microscopy. Opt. Mater. 2014;**36**(3):682–686

[44] Rocha-Mendoza I, Rangel-Rojo R, Rodríguez-Fernández L, Oliver A. Second-order nonlinear response of composites containing aligned elongated silver nanoparticles. Opt. Express. 2011;**19**(22):21575–21587

Section 4

Advanced Biological Applications

Chapter 6

Surface Modification of Silicone Rubber by Ion Implantation to Improve Biocompatibility

Xin Zhou, Yiming Zhang, Xiaohua Shi and Dongli Fan

Additional information is available at the end of the chapter

http://dx.doi.org/10.5772/intechopen.68298

Abstract

Silicone Rubber (SR) and SR-based materials have been used as medical tissue implants in the field of plastic surgery for many years, but there are still some reports of adverse reactions to long-term implants. In our study, three types of carbon ion silicone rubber were obtained by implanting three doses of carbon ions. Then, the surface characteristics, the antibacterial adhesion properties and in vivo host responses were evaluated. These study shown that ion implantation change the surface roughness and zeta potential of virgin SR; it also inhibit bacterial adhesion. At the same time, ion implantation enhance the cell proliferation, adhesion and tissue compatibility. These data indicate that carbon ion implanted silicone rubber exhibits good antibacterial adhesion properties, cytocompatibility and triggers thinner and weaker tissue capsules. In addition, according to the surface characteristics, we speculate that high surface roughness and high zeta potential may be the main factors that induce the unique biocompatibility of carbon ion implanted silicone rubber. In this chapter, we will review these results above and propose that ion implantation should be considered for further investigation and application, and carbon ion silicone rubber could be a better biomaterial to decrease silicone rubber–initiated complications.

Keywords: ion implantation, silicone rubber, surface modification, biocompatibility

1. Introduction

Silicone rubber (SR) is a type of biomaterial that exhibits many useful properties, such as thermal stability, chemical resistance, and low cost [1]. It has a long history of use in biomedical

INTECH
open science | open minds

© 2017 The Author(s). Licensee InTech. This chapter is distributed under the terms of the Creative Commons Attribution License (http://creativecommons.org/licenses/by/3.0), which permits unrestricted use, distribution, and reproduction in any medium, provided the original work is properly cited.

and biological applications, ranging from tissue fillers to tubes for dialysis and blood pumps [2–4]. Silicone rubber and silicone rubber-based materials have been used as medical tissue implants in the field of plastic surgery for many years, but there remain reports of adverse reactions to long-term implants, such as capsular contracture [5, 6]. Moreover, various prosthetic materials that contain silicone rubber can easily move and can permanently damage the prostheses. In addition, certain materials made from silicone rubber, such as catheters, are widely used in medicine but have several limitations; for example, bacteria can readily colonize the surfaces of silicone rubber, facilitating infection and even causing patient death in certain cases [5, 7–9].

In recent years, many attempts have been made to modify medical materials to reduce bacterial adhesion and to minimize adverse inflammatory or foreign body reactions [10–15]. Among these methods, the surface modification of biomaterials is an economical and effective method to achieve biocompatibility and biofunctionality while preserving the favorable characteristics of the biomaterial, such as particular mechanical properties and thermal stability. Surface modifications, such as ion implantation [16–18], sintering [19], electrochemical deposition [20], and the sol-gel coating method, are common [21]. Among these surface modification methods, ion implantation has become a notably useful method because of its ease of operation and convenience [22–25]. In this context, we implanted three different doses of carbon ions into silicone rubber and obtained carbon ion silicone rubber (C-SR). Our study was designed to evaluate the surface characteristics and the biocompatibility of carbon ion silicone rubber. We focused on bacterial adhesion, cytocompatibility, and fibrosis/fibrous capsule development. The long-term goal is to gain a better biomaterial for use by plastic surgeons.

2. Materials and methods

2.1. Sample preparation

SR sheets with dimensions of 100 x 100 x 1 mm were prepared from a two-component silicone system. Both component A and component B were clinical-quality liquids provided by Chenguang Research Institute of Chemical Engineering, China. Three doses of carbon ions were implanted using the ion implanter, respectively. The doses are 1×10^{15} ions/cm^2 (C1), 3×10^{15} ions/cm^2 (C2), and 1×10^{16} ions/cm^2 (C3). After that, SR and C-SR sheets were manufactured into disc-like samples with a diameter of 6 mm using a hole puncher and into square samples with dimensions of 10 mm × 10 mm × 1 mm. The disc-like samples were used in in vitro antibacterial adhesion tests, and the square samples were used for in vivo evaluation. All samples were sterilized with 75% alcohol overnight. In all experiments, SR served as the control.

2.2. Surface characterization

The mechanical properties were analyzed by shore A durometer and electronic universal testing machine to gain parameters of the shore hardness and tear strength. The physicochemical properties were analyzed by the FTIR, SEM, XRD, XPS, water contract angel, zeta potential, and AFM experiments. The surface zeta potentials of materials were measured with a Zeta

Potential Analyzer (DelsaNano C, Beckman Coulter, Germany). The measurements were carried out in 1 mmol 1-L NaCl electrolyte solution and with standard particles for flat surface cell (Otsuka Electronic Co., Ltd, Japan). Each sample chooses five points and each point tests 20 times. After that, the samples were scanned by the Environment Control Scanning Probe Microscope (NanoNavi E-Sweep, NSK Ltd., Tokyo, Japan) and each sample was imaged with a 5 μm × 5 μm scanned area. The surfaces were analyzed by measuring the average surface roughness (Ra) of 5 randomly chosen images per sample from selected areas of 1 μm x 1 μm under Atomic Force Microscope (AFM) analysis software (NanoNavi II, SII Nano Technology Inc., Tokyo, Japan). Ra is defined as the average absolute deviation of the roughness irregularities from the mean line over one sampling length and gives a good general description of height variations. Three replicas were used.

2.3. Bacterial culture preparation

Gram-negative *Escherichia coli* (ATCC 25922), Gram-positive *Staphylococcus aureus* (ATCC 25923), and *S. epidermidis* (ATCC 12228) were employed to bacterial experiments. The strains were streaked on blood agar plates from frozen stocks and grown for 24 h at 37°C in ambient air. The agar plates were then kept at 4°C until further use. For each experiment, one colony from an agar plate was inoculated into 10 ml of tryptone soy broth (TSB) and incubated for 24 h. The bacterial suspension was then added to 0.9% sterile sodium chloride to a final concentration of 1.5 x 10^6 colony-forming units per ml (CFU/ml), after which a McFarland standard was prepared (in practical terms, OD600 nm = 0.132). The samples were placed on 96-well culture plates and separately incubated in 200 μl of the bacterial suspension at 37°C for 1 or 24 h. After that, the plate colony-counting, fluorescence staining, and scanning electron microscopy (SEM) observation were conducted.

2.3.1. Plate colony-counting

After incubation of the samples in the bacterial culture for 1 and 24 h, the bacteria on each sample were gently rinsed with PBS, respectively, and ultrasonically detached in 1 ml of PBS solution. The bacteria in the PBS were recultivated on agar plates for colony-counting. The antibacterial rates were determined based on the following relationship: antibacterial rate (%) = (CFU of control − CFU of experimental groups)/CFU of control x 100%. This assay was performed in triplicate.

2.3.2. Fluorescence staining

After incubation for 1 and 24 h, various samples were gently rinsed with PBS before staining the bacteria on the samples. The staining was performed by applying LIVE/DEAD® BacLight™ Bacterial Viability Kit (L7029, Molecular Probes®, OR, USA) for 15 min in darkness and was examined by laser scanning confocal microscopy (LEICA TCS SP5, Germany). The areas of green and red color in the pictures were then analyzed by using Image-Pro Plus version 6.0 (Media Cybernetics, Inc., USA) and then the proportion of red coloration based on the following relationship was calculated: red proportion (%) = red area/(green area + red area) x 100%.

2.3.3. Scanning electron microscopy (SEM) observation

After bacterial incubation for 1 and 24 h, the samples were rinsed with PBS to remove free bacterial cells and then fixed in 2.5% glutaraldehyde for 3 h at room temperature. The samples were then progressively dehydrated in a series of ethanol solutions (15, 30, 50, 70, 80, 90, 95, and 100%) for 15 min each. After that, the specimens underwent critical point drying and coating with a thin conductive layer of Au. Finally, the morphology and adhesion of the bacteria on the various samples were determined by SEM (VEGA 2 SEM, TESCAN Inc., Brno, Czech Republic).

2.4. Animals and surgery

This research was performed in accordance with the Guide for the Care and Use of Laboratory Animals published by the US National Institutes of Health (Washington, DC: The National Academies Press, 2011), and all of the animal protocols were approved by the Institutional Animal Care and Use Committee of the Third Military Medical University, China. A total of 16 female Sprague-Dawley rats (weighing 160–200 g) were used (4 groups of 4 animals each), and all rats were housed under a 12-h light–dark cycle with free access to water and food. Prior to surgery, all of the rats were anesthetized with 3% pentobarbital sodium (1 ml/1000 g). The skin was swabbed with iodine, and four parallel incisions (10 mm) were performed. The material samples were implanted subcutaneously along the back region. The implants and their surrounding tissues were retrieved from each group by wide excision at 7, 30, 90, and 180 days after implantation and were then fixed in 4% paraformaldehyde solution. After that, HE and Masson's staining and immunohistochemistry were carried out.

2.4.1. HE and Masson's staining

The fixed tissues were sectioned (6 μm thick) and stained using a HE Staining Kit (C0105, Beyotime Inc., Shanghai, China). The thickness of the fibrotic capsule around each implant was determined at five equidistant points for statistical accuracy. Collagen deposition in the tissue around the implants was studied by Masson's trichrome staining, which was performed using a staining kit (MST-8003, Maixin Biological Technology Co., Ltd., Fujian, China). All procedures were performed based on the manufacturer's instructions.

2.4.2. Immunohistochemistry

Immunohistochemistry was performed on 4% paraformaldehyde-fixed cryostat sections of frozen tissue specimens. Endogenous peroxidase and nonspecific antibody binding were blocked with 3% H_2O_2 and 100% methanol, with a ratio of 1:5 and a blocking time of 30 min. Next, 0.02 M PBS was used for antigen retrieval while heating in a water bath, followed by treatment with 5% blocking reagent at 37°C for 30 min. The slides were then incubated at 4°C for 12 h with a primary antibody against CD68, CD4, TNF-α, α-SMA, or elastin (1:25) (Boster Biological Engineering Co., Ltd., Hubei, China). After washing in PBS, a secondary antibody was applied for 30 min. Visualization was achieved by adding 3, 3′-diaminobenzidine chromogen.

2.5. Statistics

All data are expressed as the mean ± standard deviation (SD), and statistics were analyzed using SPSS statistical software. One-way ANOVA combined with multiple comparisons performed along with Tukey multiple comparison tests was utilized to determine the level of significance. In all of the statistical evaluations, $P < 0.05$ was considered significant.

3. Results

3.1. Ion implantation changes the surface roughness and zeta potential of SR

After ion implantation, SEM, AFM, FTIR, XPS, XRD, water contact angle measure instrument, zeta potential detection instrument, shore A durometer, and an electronic universal testing machine were used to investigate the change in properties of carbon ion silicone rubber. The SEM results failed to find any significant differences between virgin SR and three C-SRs (**Figure 1**), indicating that carbon ion implantation did not change the macro-scale surface of SR.

At the same time, there is no any significant differences or the difference was very small on the results of FTIR (**Figure 2**), XRD (**Figure 3**), shore hardness (**Figure 4A**), and tear strength (**Figure 4B**).

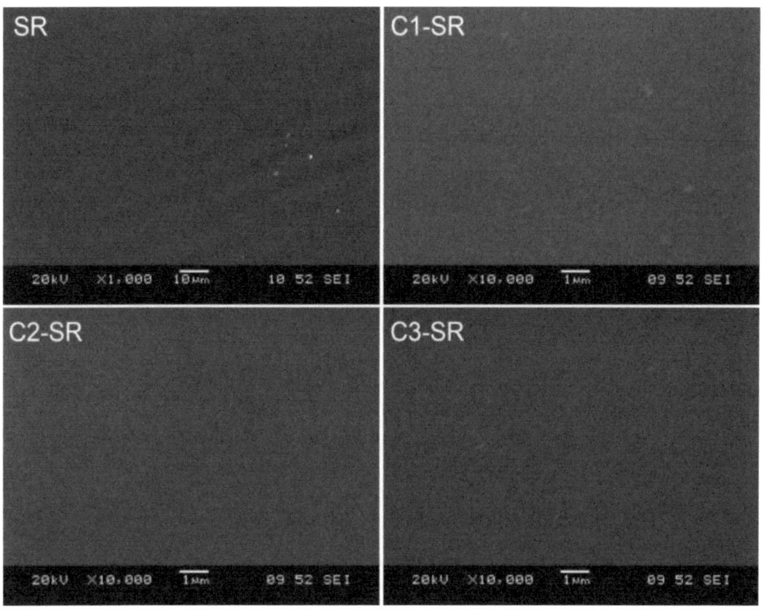

Figure 1. Representative scanning electron microscopic images of virgin silicone rubber and carbon ion-implanted silicone rubber.

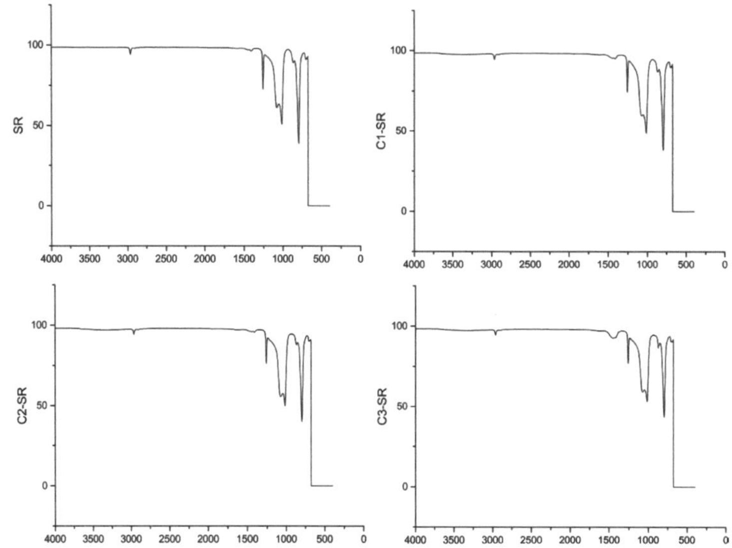

Figure 2. The Fourier transform infrared spectroscopy results of virgin silicone rubber and carbon ion-implanted silicone rubber.

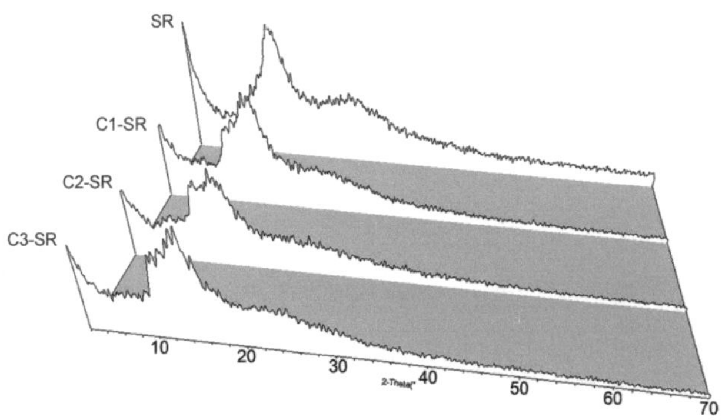

Figure 3. The XRD results of virgin silicone rubber and carbon ion-implanted silicone rubber.

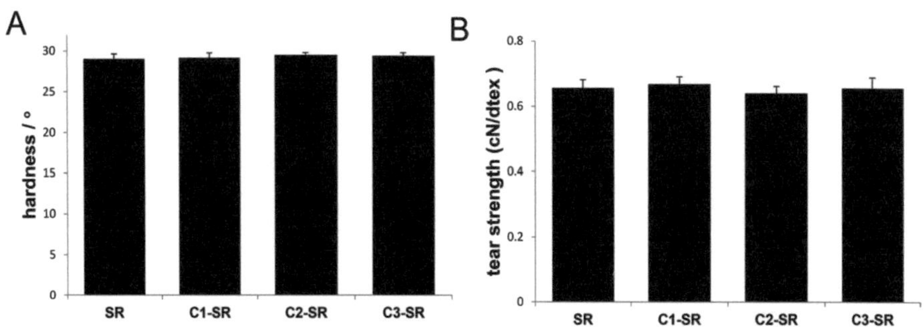

Figure 4. The results of Shore A hardness and tear strength of virgin silicone rubber and carbon ion-implanted silicone rubber. A, Shore A hardness. B, Tear strength.

From the results of water contact angle, we found that carbon ion implantation significantly decreased the water contact angle of SR, and big doses carbon ions had lowest water contact angle among all C-SRs (**Figure 5**).

Besides, the XPS results showed that carbon ion implantation significantly changed the surface silicone oxygen rate and chemical-element distribution of SR (**Figure 6**) (**Table 1**); we noted

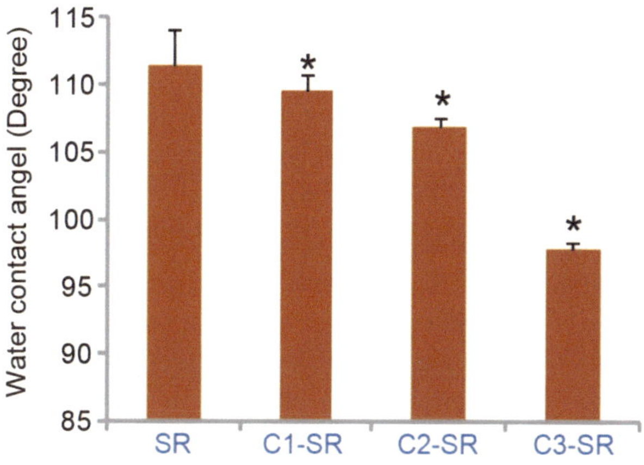

Figure 5. Water contact angle of virgin silicone rubber and carbon ion-implanted silicone rubber (*$P < 0.05$ compared with silicone rubber).

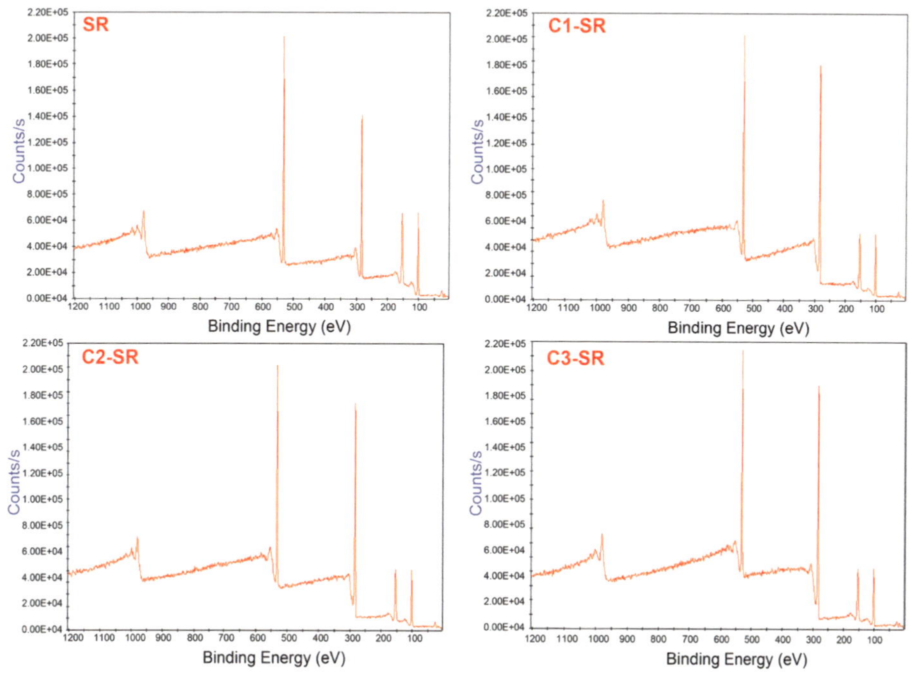

Figure 6. XPS results of virgin silicone rubber and carbon ion-implanted silicone rubber.

Group	Si 2p	C 1s	O 1s
SR	28.82	47.57	23.65
C1-SR	18.94	58.40	22.67
C2-SR	18.96	58.44	22.60
C3-SR	18.13	61.49	20.38

Table 1. Chemical composition (in at.%) from the XPS analysis.

that with the ion implantation dose increasing, the carbon content in the material increased, while the Si content decreased, suggesting that implanted carbon atom may replace the Si of virgin SR, interrupting the original Si-O assemble, increasing the surface free energy, and, thereby, theoretically decreasing material's water contact angle.

Furthermore, AFM images revealed that the surfaces of C-SRs were composed of larger irregular peaks and deeper valleys, while virgin SR exhibited a relatively smooth and more homogeneous surface (**Figure 7A**). The surface roughness of the C3-SR, which underwent most carbon ion implantation, was highest among all three C-SRs (**Figure 7B**).

In addition, all samples exhibited negative zeta potentials and reflect that the surfaces of all samples were negatively charged. The absolute value of the zeta potential increased with the ion dose (**Figure 8**). Considering the influence of surface roughness on contact angle, we propose that ion implantation can change the surface roughness of the material and increase the surface potential of the material.

3.2. Ion implantation inhibits bacterial adhesion on SR

Preventing bacterial adhesion and biofilm formation by improving the surface antibacterial adhesion property of the silicone rubber is critical for eliminating various types of infections.

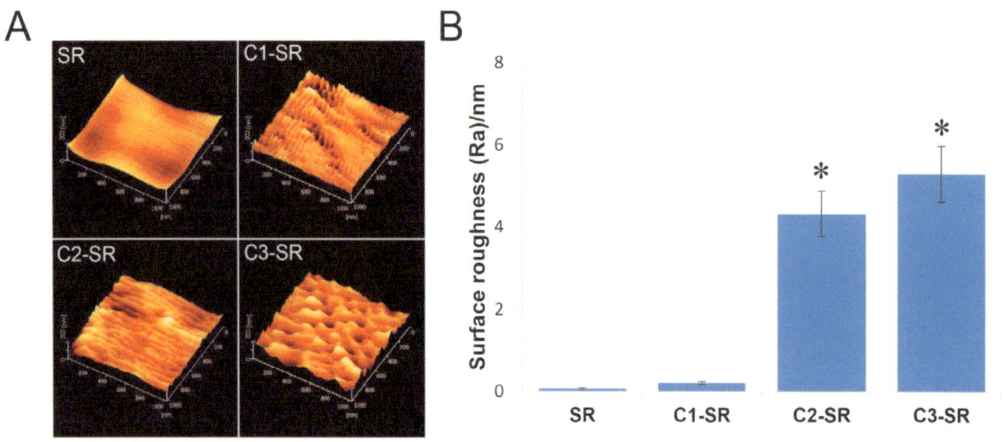

Figure 7. AFM results of virgin silicone rubber and carbon ion-implanted silicone rubber. A, Representative atomic force microscope images. B, Surface roughness (*P < 0.01 compared with silicone rubber).

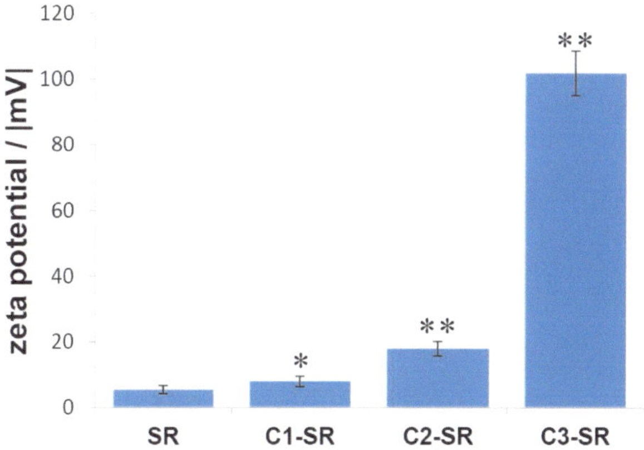

Figure 8. The zeta potential of virgin silicone rubber and carbon ion-implanted silicone rubber (*P < 0.05 compared with silicone rubber; **P < 0.01 compared with silicone rubber).

After ion implantation, we used Gram-negative *Escherichia coli* (American Type Culture Collection 25922) to evaluate the ability to resist bacteria adhesion. From the result, after 1 h of incubation, the rate of *E. coli* adherence on the carbon ion silicone rubber (1×10^{15} carbon ions/cm^2; 3×10^{15} carbon ions/cm^2; and 1×10^{16} carbon ions/cm^2) increased to approximately 11, 25, and 33%, respectively, and that on carbon ion silicone rubber increased significantly after 1 h of incubation (**Figure 9**) ($P < 0.05$).

After 24 h of incubation, the rates of bacterial adherence were slightly lower, but did not significantly decrease compared with that after 1 h of incubation ($P > 0.05$). The ability of carbon ion silicone rubber to prevent viable bacteria colonization was also verified by fluorescence staining. The results had shown that the amount of bacterial adhesion to the surface of carbon ion silicone rubber was reduced compared with the virgin silicone rubber (**Figure 10**). Scanning electron microscopy was performed to examine the attached bacteria. As the results

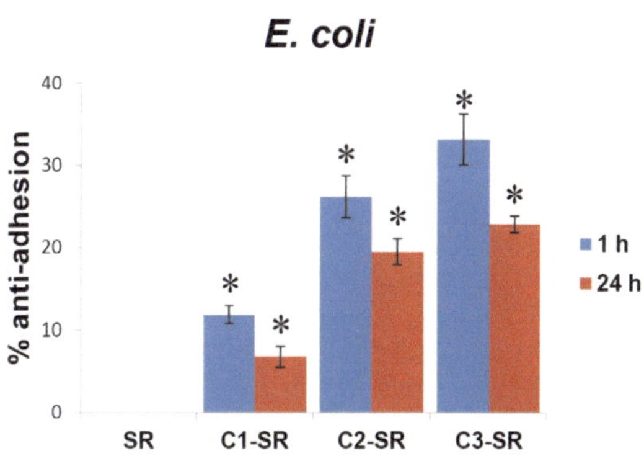

Figure 9. The antiadhesion rates (%) of virgin silicone rubber and carbon ion-implanted silicone rubber. After all samples were cultured in bacterial suspension for 1 and 24 h, bacteria on the surface of all samples were recultured on the plate, and bacterial colonies were subsequently counted. According to the number of colonies, the antiadhesion rates (%) for *E. coli* were calculated. The data are presented as the mean ± SD (n = 3); *$P < 0.05$ compared with silicone rubber.

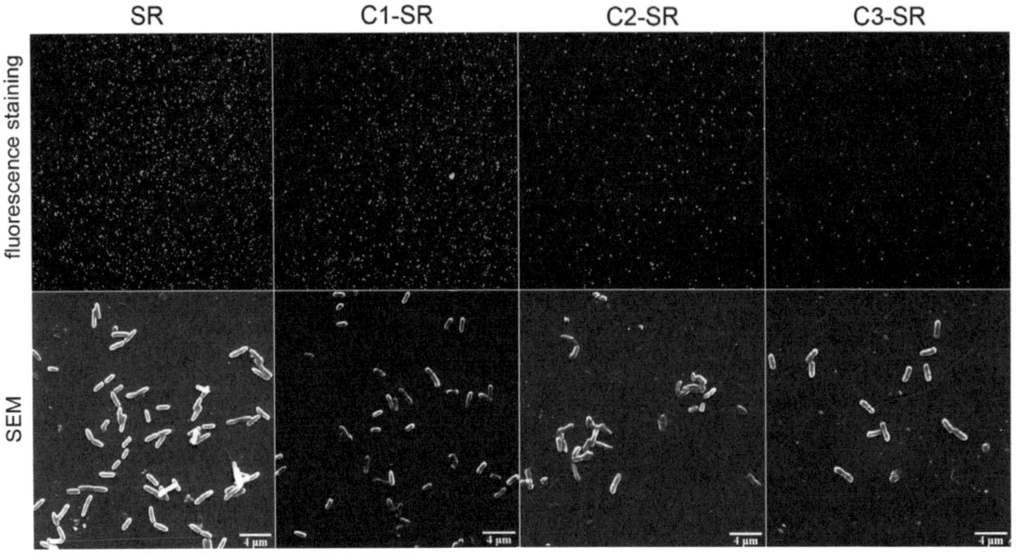

Figure 10. Representative images of fluorescence staining and scanning electron microscopy observation of virgin silicone rubber and carbon ion-implanted silicone rubber. Representative images showing bacteria viability on SR and CSR after 24 h of incubation, as indicated by staining with a LIVE/DEAD BacLight Bacterial Viability Kit (Thermo Fisher Scientific, Waltham, Mass). The live bacteria appear green, whereas the dead bacteria are red (original magnification, × 200). Representative scanning electron microscopic images of the bacteria on SR and CSR after incubation for 24 h.

have shown that bacteria were observed on surfaces of all samples, but there were differences in quantity (**Figure 10**).

3.3. Carbon ion-implanted silicone rubber triggers thinner and weaker tissue capsules

After ion implantation, the host responses were evaluated by surveying inflammation and fibber capsule formation that developed after subcutaneous implantation in Sprague-Dawley rats for 7, 30, 90, and 180 days. The thickness values of tissue capsules around the implants were identified from hematoxylin and eosin-stained sections of the peri-implant soft tissues and were analyzed as one of the physiologic responses to implantation. At 7 days after implantation, silicone rubber had the thinnest tissue capsules, and carbon ion silicone rubber had thicker (**Figure 11**) ($P > 0.05$) and weaker tissue capsules. Interestingly, the thickness decreased with longer implantation and increasing carbon ion doses (**Figure 11**). At 180 days after implantation, silicone rubber and C3-SR had the thickest and the thinnest tissue capsules, respectively (**Figure 11**).

In addition, collagen deposition was revealed using Masson trichrome staining. Our results show that collagen gradually became sparser over time and with increasing carbon ion doses. Carbon ion silicone rubber had obviously lower collagen deposition than silicone rubber (**Figure 12**) ($P < 0.05$).

To gain insight into inflammatory foreign body responses and capsule contracture to the samples, major biomarkers CD68, CD4, tumor necrosis factor-α, elastin, and α-smooth muscle

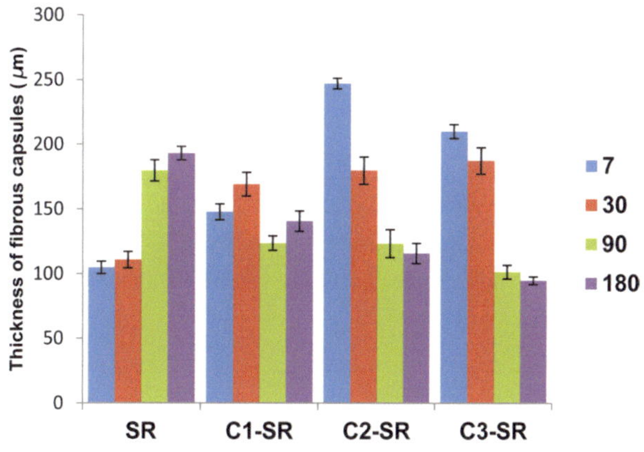

Figure 11. The capsule thicknesses around the implants.

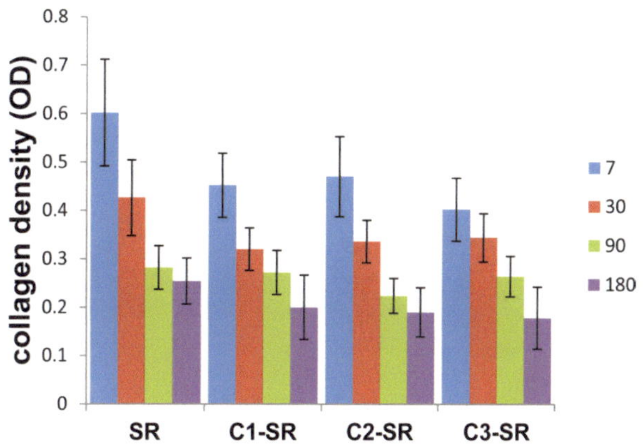

Figure 12. The collagen density around the implants.

actin were detected using immunohistochemistry. As shown in **Table 2**, all samples present lower expression of CD68, with no significant differences.

The distribution of CD4 in the inflammatory infiltrate, which was induced by the samples, was investigated to further understand the local immunomodulation against these types of materials. The results show that there were many positive staining areas of CD4 in silicone

	7 days	30 days	90 days	180 days
SR	+	+	+	+
C1-SR	+	+	+	+
C2-SR	+	+	+	+
C3-SR	+	+	+	+

Table 2. Semiquantitative evaluation of CD68 in peri-implant tissue.

rubber after 90 days, but positive staining in the carbon ion silicone rubber decreased with time. After 90 days, CD4 significantly decreased compared with silicone rubber (**Table 3**).

In addition, the expression results of proinflammatory cytokine tumor necrosis factor-α by macrophage cells show that silicone rubber had an obviously positive staining area (**Table 4**).

Furthermore, the positive staining areas of α-smooth muscle actin and elastin have no difference; the positive staining area of α-smooth muscle actin appeared predominantly in silicone rubber than in carbon ion silicone rubber (**Table 5**). Elastin was intensely expressed in silicone rubber, particularly after 30 days (**Table 6**).

	7 days	30 days	90 days	180 days
SR	++	++	+++	+++
C1-SR	+++	++	+	+
C2-SR	++	++	++	++
C3-SR	++	+	+	+

Table 3. Semiquantitative evaluation of CD4 in peri-implant tissue.

	7 days	30 days	90 days	180 days
SR	++	++	+++	+++
C1-SR	++	++	+	+
C2-SR	+++	+	+	+
C3-SR	++	+	+	+

Table 4. Semiquantitative evaluation of TNF-α in peri-implant tissue.

	7 days	30 days	90 days	180 days
SR	+	++	++	++
C1-SR	+	+	+	+
C2-SR	+	+	+	+
C3-SR	+	+	+	+

Table 5. Semiquantitative evaluation of α-SMA in peri-implant tissue.

	7 days	30 days	90 days	180 days
SR	++	+++	+++	++
C1-SR	+	++	++	+
C2-SR	++	+	++	+
C3-SR	++	++	+	+

Table 6. Semiquantitative evaluation of elastin in peri-implant tissue.

4. Conclusion

Our study evaluated the in vitro antibacterial properties and the in vivo host response to carbon ion-implanted silicone rubber. The results of our study indicate that the carbon ion silicone rubbers have good biocompatibility, lower bacterial adhesion, and lower foreign body reaction with relatively thin fibrous capsules. All results show that ion implantation should be considered for further investigation and application, and carbon ion silicone rubber might be a better biomaterial for decreasing silicone rubber-initiated complications.

Acknowledgements

This work was funded by grants from the National Natural Science Foundation of China (81571918, 81401610) and a grant for Scientific Personnel Innovation from Chongqing (cstc2013kjrc-qnrc10003).

Author details

Xin Zhou, Yiming Zhang, Xiaohua Shi and Dongli Fan*

*Address all correspondence to: fdltmmu@sina.com

Department of Plastic and Cosmetic Surgery, Xinqiao Hospital, Third Military Medical University, Chongqing, China

References

[1] Liu P, Chen Q, Yuan B, Chen M, Wu S, Lin S, Shen J. Facile surface modification of silicone rubber with zwitterionic polymers for improving blood compatibility. Materials Science & Engineering C-Materials for Biological Applications. 2013;**33**(7):3865-3874. DOI: 10.1016/j.msec.2013.05.025

[2] Lugowski SJ, Smith DC, Bonek H, Lugowski J, Peters W, Semple J. Analysis of silicon in human tissues with special reference to silicone breast implants. Journal of Trace Elements in Medicine and Biology. 2000;**14**(1):31-42. DOI: 10.1016/S0946-672X(00)80021-8

[3] Erlich MA, Parhiscar A. Nasal dorsal augmentation with silicone implants. Facial Plastic Surgery. 2003;**19**(4):325-330. DOI: 10.1055/s-2004-815652

[4] Cazacu M, Racles C, Vlad A, Antohe M, Forna N. Silicone-based composite for relining of removable dental prosthesis. Journal of Composite Materials. 2009;**43**(10):2045-2055. DOI: 10.1177/0021998309340447

[5] Anderson JM, Rodriguez A, Chang DT. Foreign body reaction to biomaterials. Seminars in Immunology. 2008;**20**(2):86-100. DOI: 10.1016/j.smim.2007.11.004

[6] Rohrich RJ. Advances in breast augmentation. Plastic and Reconstructive Surgery. 2006;**118**(7):1S-2S. DOI: 10.1097/01.prs.0000247292.96867.06

[7] Vasilev K, Cook J, Griesser HJ. Antibacterial surfaces for biomedical devices. Expert Review of Medical Devices. 2009;**6**(5):553-567. DOI: 10.1586/erd.09.36

[8] Bazaka K, Jacob MV, Crawford RJ, Ivanova EP. Plasma-assisted surface modification of organic biopolymers to prevent bacterial attachment. Acta Biomaterialia. 2011; **7**(5):2015-2028. DOI: 10.1016/j.actbio.2010.12.024

[9] Bergmann PA, Tamouridis G, Lohmeyer JA, Mauss KL, Becker B, Knobloch J, Mail Nder P, Siemers F. The effect of a bacterial contamination on the formation of capsular contracture with polyurethane breast implants in comparison with textured silicone implants: An animal study. Journal of Plastic Reconstructive and Aesthetic Surgery. 2014;**67**(10):1364-1370. DOI: 10.1016/j.bjps.2014.05.040

[10] Tang CY, Chen D-Z, Chan KYY, Chu KM, Ng PC, Yue TM. Fabrication of antibacterial silicone composite by an antibacterial agent deposition, solution casting and crosslinking technique. Polymer International. 2011;**60**(10):1461-1466. DOI: 10.1002/pi.3102

[11] Xue L, Zhang Y, Zuo Y, Diao S, Zhang J, Feng S. Preparation and characterization of novel UV-curing silicone rubber via thiol-ene reaction. Materials Letters. 2013;**106**:425-427. DOI: 10.1016/j.matlet.2013.05.084

[12] van der Houwen EB, Kuiper LH, Burgerhof JG, van der Laan BF, Verkerke GJ. Functional buckling behavior of silicone rubber shells for biomedical use. Journal of the Mechanical Behavior of Biomedical Materials. 2013;**28**:47-54. DOI: 10.1016/j.jmbbm.2013.07.002

[13] Abbasi F, Mirzadeh H, Katbab AA. Modification of polysiloxane polymers for biomedical applications: A review. Polymer International. 2001;**50**(12):1279-1287. DOI: 10.1002/pi.783

[14] Magennis EP, Hook AL, Williams P, Alexander MR. Making silicone rubber highly resistant to bacterial attachment using thiol-ene grafting. ACS Applied Materials & Interfaces. 2016;**8**(45):30780-30787. DOI: 10.1021/acsami.6b10986

[15] Ziraki S, Zebarjad SM, Hadianfard MJ. A study on the role of polypropylene fibers and silica nanoparticles on the compression properties of silicone rubber composites as a material of finger joint implant. International Journal of Polymeric Materials and Polymeric Biomaterials. 2017;**66**(1):48-52. DOI: 10.1080/00914037.2016.1180623

[16] Tsuji H, Izukawa M, Ikeguchi R, Kakinoki R, Sato H, Gotoh Y, Ishikawa J. Surface treatment of silicone rubber by carbon negative-ion implantation for nerve regeneration. Applied Surface Science. 2004;**235**(1–2):182-187. DOI: 10.1016/j.apsusc.2004.05.121

[17] Tsuji H, Sommani P, Hattori M, Yamada T, Sato H, Gotoh Y, Ishikawa J. Negative ion implantation for patterning mesenchymal-stem cell adhesion on silicone rubber and

differentiation into nerve cells with keeping their adhesion pattern. Surface and Coatings Technology. 2009;**2003**(17–18):2562-2565. DOI: 10.1016/j.surfcoat.2009.02.076

[18] Tsuji H, Sommani P, Kitamura T, Hattori M, Sato H, Gotoh Y, Ishikawa J. Nerve-cell attachment properties of polystyrene and silicone rubber modified by carbon negative-ionimplantation. Surface and Coatings Technology. 2007;**201**(19–10):8123-8126. DOI: 10.1016/j.surfcoat.2006.01.074

[19] Rahman CV, Ben-David D, Dhillon A, Kuhn G, Gould TWA, Muller R, Rose F, Shakesheff KM, Livne E. Controlled release of BMP-2 from a sintered polymer scaffold enhances bone repair in a mouse calvarial defect model. Journal of Tissue Engineering and Regenerative Medicine. 2014;**8**(1):59-66. DOI: 10.1002/term.1497

[20] Ling T, Lin J, Tu JJ, Liu SQ, Weng WJ, Cheng K, Wang HM, Du PY, Han GR. Mineralized collagen coatings formed by electrochemical deposition. Journal of Materials Science-Materials in Medicine. 2013;**24**(12):2709-2718. DOI: 10.1007/s10856-013-5028-9

[21] Meretoja VV, Rossi S, Peltola T, Pelliniemi LJ, Narhi TO. Adhesion and proliferation of human fibroblasts on sol–gel coated titania. Journal of Biomedical Materials Research Part A. 2010;**95A**(1):269-275. DOI: 10.1002/jbm.a.32829

[22] Baba K, Hatada R, Flege S, Ensinger W. Preparation and properties of Ag-containing diamond-like carbon films by magnetron plasma source ion implantation. Advances in Materials Science and Engineering. 2012;536853. DOI: 10.1155/2012/536853

[23] Dudognon J, Vayer M, Pineau A, Erre R. Mo and Ag ion implantation in austenitic, ferritic and duplex stainless steels: A comparative study. Surface and Coatings Technology. 2008;**203**(1–2):180-185. DOI: 10.1016/j.surfcoat.2008.08.069

[24] Fang J, Zhao JH, Sun Y, Ma HY, Yu XL, Ma Y, Ni YX, Zheng L, Zhou YM. Biocompatibility and antibacterial properties of zinc-ion implantation on titanium. Journal of Hard Tissue Biology. 2014;**23**(1):35-43.

[25] Hou XG, Wang XM, Luan WJ, Li DJ, Dong L, Ma J. Study of antibacterial, hydrophilic and nanomechanical properties of TiOx films modified by Ag+ beam implantation. Surface and Coatings Technology. 2013;**229**:71-75. DOI: 10.1016/j.surfcoat.2012.04.092